MECHANICS FUNDAMENTALS

Other books in the FUNdamental Series
Light FUNdamentals
Electricity and Magnetism FUNdamentals
Heat FUNdamentals
Sound FUNdamentals

MECHANICS FUNDAMENTALS

FUNtastic Science Activities for Kids

Robert W. Wood

Illustrated by Bill Wright

LEARNING
TRIANGLE
PRESS

Connecting
kids, parents, and teachers
through learning

An imprint of McGraw-Hill

New York San Francisco Washington, D.C. Auckland Bogotá Caracas
Lisbon London Madrid Mexico City Milan Montreal New Delhi
San Juan Singapore Sydney Tokyo Toronto

McGraw-Hill

A Division of The McGraw·Hill Companies

pbk   1 2 3 4 5 6 7 8 9 FGR/FGR 9 0 0 9 8 7 6
hc   1 2 3 4 5 6 7 8 9 FGR/FGR 9 0 0 9 8 7 6

Product or brand names used in this book may be trade names or trademarks. Where we believe that there may be proprietary claims to such trade names or trademarks, the name has been used with an initial capital or it has been capitalized in the style used by the name claimant. Regardless of the capitalization used, all such names have been used in an editorial manner without any intent to convey endorsement of or other affiliation with the name claimant. Neither the author nor the publisher intends to express any judgment as to the validity or legal status of any such proprietary claims.

Library of Congress Cataloging-in-Publication Data
Wood, Robert W.
    Mechanics FUNdamentals : FUNtastic science activities for kids /
Robert W. Wood ; illustrated by Bill Wright.
        p. cm.
    Includes index.
    Summary: Provides instructions for a variety of experiments to
study the effects of forces on bodies or fluids at rest or in
motion.
    ISBN 0-07-071806-7. — ISBN 0-07-071807-5 (pbk.)
    1. Mechanics—Experiments—Juvenile literature. 2. Fluid
mechanics—Experiments—Juvenile literature. 3. Dynamics—
Experiments—Juvenile literature. 4. Statics—Experiments—
Juvenile literature. [1. Mechanics—Experiments.
2. Experiments.] I. Wright, Bill, ill. II. Title.
QC127.4.W64 1996
531'.078—dc20
                                                        96-41802
                                                          CIP
                                                          AC

McGraw-Hill books are available at special quantity discounts to use as premiums and sales promotions, or for use in corporate training programs. For more information, please write to the Director of Special Sales, McGraw-Hill, 11 West 19th Street, New York, NY 10011. Or contact your local bookstore.

Acquisitions editor: Kim Tabor
Editorial team: Managing Editor: Susan Kagey
                  Book editor: Joanne M. Slike
                  Technical reviewer: Andrea T. Bennett
Production team: DTP supervisor: Pat Caruso
                  DTP operators: Tanya Howden, Kim Sheran, John Lovell
                  DTP computer artist supervisor: Tess Raynor
                  DTP computer artists: Nora Ananos, Charles Burkhour, Steven Jay Gellert,
                     Charles Nappa
                  Indexer: Jodi L. Tyler
Designer: Jaclyn J. Boone

CONTENTS

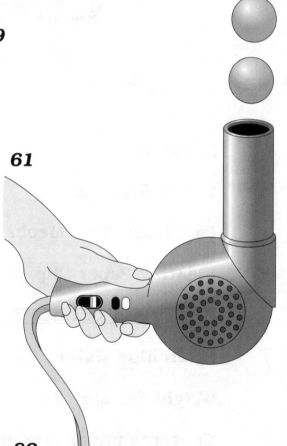

INTRO

Physics is the science that explores the natural world around us. It tells us how and why a lever can lift a heavy weight, why hot air rises, and what light is. It is the study of electricity and magnetism, and of how sound waves travel. This fascinating science covers such a wide range of subjects that no simple definition explains it.

To make physics easier to study, it is broken up into smaller fields: mechanics, heat, light, electricity and magnetism, and sound. But sometimes these fields overlap each other. For example, a student studying electricity and magnetism to learn why a telephone works would also learn how sound waves vibrate objects to send electrical signals.

This book deals with *mechanics*, or the study of the effects of forces on bodies or fluids at rest or in motion. The British mathematician and physicist Sir Isaac Newton first described building and using machines as "mechanics." Mechanics is used to design airplanes, rockets, and space vehicles. Engineers use mechanics to make sure bridges stand up under the stress of the loads they will carry. Scientists use mechanics to study the motion of atomic particles. Astronomers use the principles of mechanics to determine the movement of planets and stars, and

certainly, it's important to the success of orbiting satellites and the space shuttle.

Mechanics FUNdamentals is divided into two parts. Part One, Fluid Mechanics, deals with hydraulics, hydrostatics, and aerodynamics. In this section, you'll learn that the surface of water is surprisingly strong. You'll also learn how to determine specific gravity. The difference between weight and mass, and between speed, velocity, and acceleration is explained, as well.

Part Two, Solid Mechanics, deals with dynamics and statics. Here you'll learn about kinetic and potential energy, friction, and centrifugal and centripetal force, as well as about inertia and momentum. In addition, you'll learn how the pendulum of a clock works and how gears work. You'll also build simple machines such as ramps, pulleys, and levers.

Demonstrating the principles of physics does not require complex experiments and fancy equipment. Though the experiments contained in this book are simple to perform, they illustrate important principles about the natural world and answer many questions about why objects move or react in a certain manner. Each experiment begins with an objective, followed by a materials list and step-by-step procedures. Results are given to explain what is being demonstrated, as well as a few questions to discuss further. The experiments conclude with fun facts.

Where measurements are used, they are given in both the English and metric systems as numbers that will make the experiments easy to perform. Use whichever system you like, but realize that the numbers might not be exact equivalents.

Be sure to read Safety Stuff before you begin any experiment. It recommends safety precautions you should take. It also tells you whether you should have a teacher or another adult help you. Keep safety in mind, and you will have a rewarding first experience in the exciting world of physics.

x

SAFETY STUFF

Science experiments can be fun and exciting, but safety should always be considered. Parents and teachers are encouraged to participate with their children and students.

 Look over the steps before beginning any experiment. You will notice that some steps are preceded by a caution symbol like the one next to this paragraph. This symbol means that you should use extra safety precautions or that the experiment requires adult supervision.

Materials or tools used in some experiments could be dangerous in young hands. Adult supervision is recommended whenever the caution symbol appears. Children need to be taught about the care and handling of sharp tools or combustible or toxic materials and how to protect surfaces. Also, extreme caution must be exercised around any open flame.

Use common sense and make safety the priority, and you will have a safe and fun experience!

Part One
Fluid
Mechanics

Fluid mechanics shows how forces and motions affect fluids and gases. It includes fluids at rest, called *hydrostatics*; fluids in motion, called *hydraulics*; and air moving around objects, called *aerodynamics*. Engineers use fluid mechanics to understand the flight of aircraft, including the space shuttle; how ships move in water; the flow of water over a dam; and even the movement of atmospheric and ocean currents. The following experiments provide the basics for understanding fluid mechanics.

Break the tension
with this experiment!

FLOATING PAPER CLIP

YOUR CHALLENGE

To observe the unique strength of water's surface.

DO THIS

1   Using the fork, carefully lower the paper clip to the
    surface of the water. Slowly remove the fork. What
    happens? (Figure 1-1)

YOU NEED

Bowl of water

Fork

Paper clip

Liquid dish detergent

Sheet of waxed paper

**Clear drinking glass
of water**

Place the paper
clip on the surface
of the water.

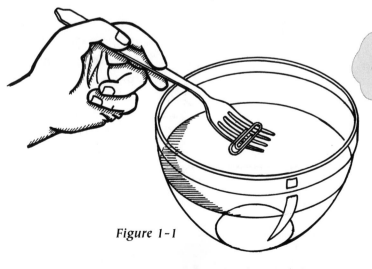

Figure 1-1

3

2   Now pour a drop of detergent into the
    water and lower the paper clip again.
    What happens now? (Figure 1-2)

Drop liquid detergent
on the water.

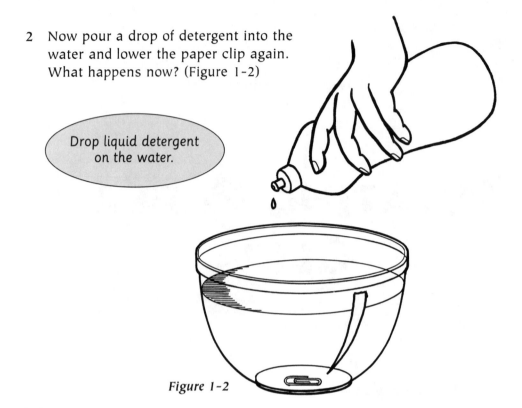

Figure 1-2

3   Sprinkle a few drops of water on a sheet of waxed paper. What shape
    do they form? Why do you think this happens? (Figure 1-3)

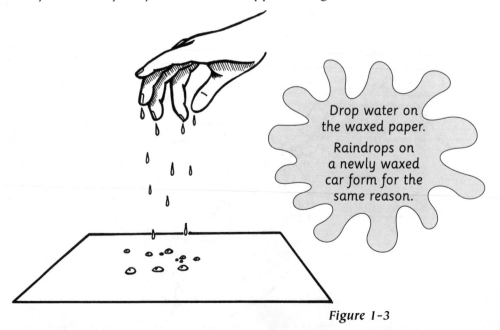

Drop water on
the waxed paper.

Raindrops on
a newly waxed
car form for the
same reason.

Figure 1-3

4 Tilt the sheet so that some of the drops run together. What happens when one drop touches another? Now what shape do they form? (Figure 1-4)

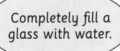

Run the drops together.

Figure 1-4

5 Fill a glass to the rim with water. Now add a few more drops of water until it almost overflows. Look at the surface of the water from the side. What is the shape of the surface? Why do you think it has this shape? (Figure 1-5)

Figure 1-5

Completely fill a glass with water.

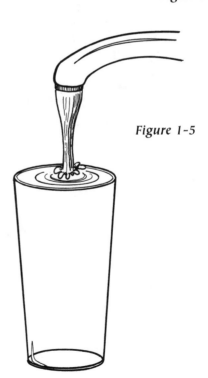

WHAT HAPPENED?

The surface of water, whether it is flat or in the form of a round drop, has a thin skin called *surface tension*. The molecules in water have a strong attraction for each other. They pull toward each other in all directions. But at the surface, there are no other molecules to pull to, so they press together, forming a thin skin. The skin is surprisingly strong and will support objects that would normally sink. The detergent lowers the surface tension and the skin becomes weaker.

How do small water droplets in a cloud form larger raindrops?

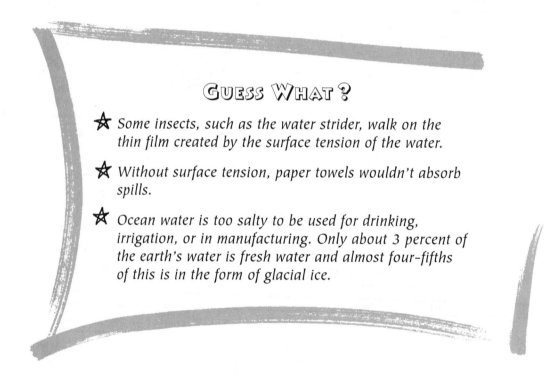

GUESS WHAT?

★ *Some insects, such as the water strider, walk on the thin film created by the surface tension of the water.*

★ *Without surface tension, paper towels wouldn't absorb spills.*

★ *Ocean water is too salty to be used for drinking, irrigation, or in manufacturing. Only about 3 percent of the earth's water is fresh water and almost four-fifths of this is in the form of glacial ice.*

Roil some oil!

GREASY KID'S STUFF

YOUR CHALLENGE

To observe the reaction between oil and surface tension.

DO THIS

1   Fill a drinking glass two-thirds full with water. Add a few drops of oil to the surface of the water. What do you see? (Figure 2-1)

YOU NEED

Drinking glass

Water

Cooking oil

Soap

Toothpick

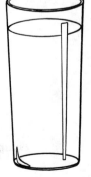

Add a couple of drops of cooking oil to the water.

Figure 2-1

7

2 Use the toothpick to try to move
   the oil spot over the surface of
   the water. What happens?
   Can you guess why?
   (Figure 2-2)

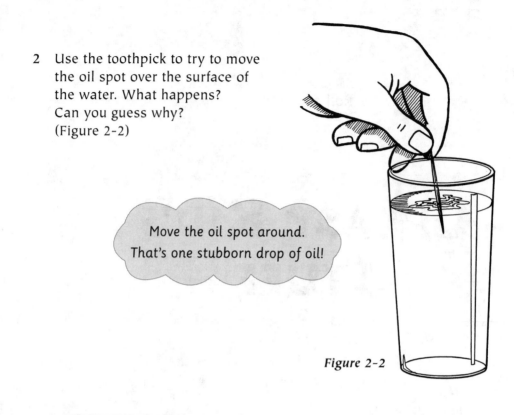

Move the oil spot around.
That's one stubborn drop of oil!

Figure 2-2

Add soap to the tip of
the toothpick.

Did you know that oil
spills in the ocean
are cleaned up
with detergent?

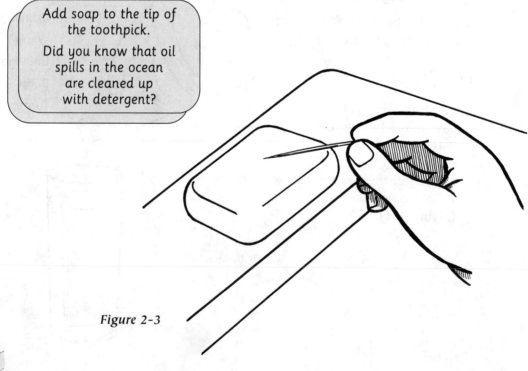

Figure 2-3

> Touch the soap to the oil spot.
>
> The soap just made the oil scram!

Figure 2-4

3   Place a small amount of soap on the tip of the toothpick and gently touch it to the center of the oil spot. What happens now? (Figures 2-3 and 2-4)

Why did the oil form a round spot? Could you move the oil spot with the toothpick? Where does the oil spot want to stay? Why do you think the spot behaves this way?

WHAT HAPPENED?

When you touched the center of the oil spot with soap, the soap lowered the surface tension of the water below the center of the oil drop. This means that the surface tension of the water outside the drop was stronger than in the center. The stronger surface tension pulled the oil in an outward direction, causing it to instantly spread in a wide circle to the sides of the glass. Soap lowers the surface tension of water, but the water still has some surface tension.

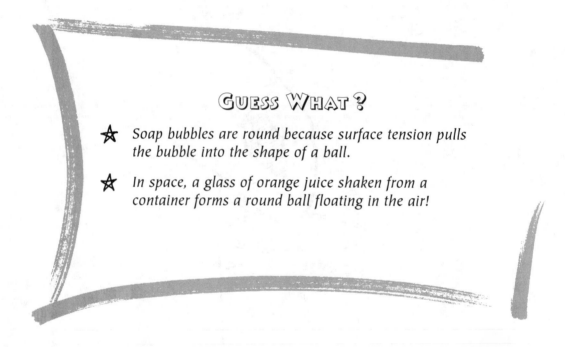

GUESS WHAT?

★ Soap bubbles are round because surface tension pulls the bubble into the shape of a ball.

★ In space, a glass of orange juice shaken from a container forms a round ball floating in the air!

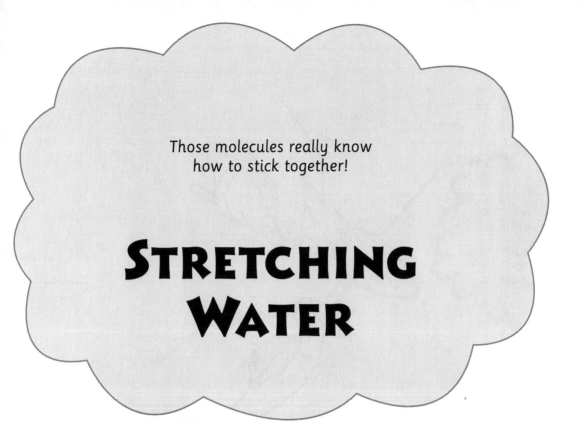

Those molecules really know how to stick together!

STRETCHING WATER

YOUR CHALLENGE

To observe the pulling power of surface tension.

DO THIS

1 Carefully cut a straight section of wire about 4 inches long, or have an adult do it. (Figure 3-1)

2 Bend the remaining wire into a long, narrow loop with a twisted handle sticking up. (Figure 3-2)

3 Fill the sink with just enough soapy water to cover the loop when it is lowered.

4 Now place the straight wire across the middle of the loop. Keeping the loop level, briefly lower it into the water and slowly raise it back up. Your loop should now have a thin film of water covering it and the straight wire. (Figure 3-3)

YOU NEED

Smooth wire (such as a thin coat hanger)

Pliers with wire cutters

Kitchen sink

Soapy water

11

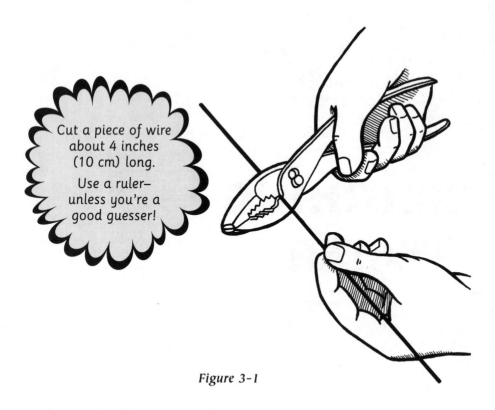

Cut a piece of wire about 4 inches (10 cm) long.

Use a ruler–unless you're a good guesser!

Figure 3-1

Twist the wires to make a handle.

Make sure to use wire that's easy to bend.

Figure 3-2

12

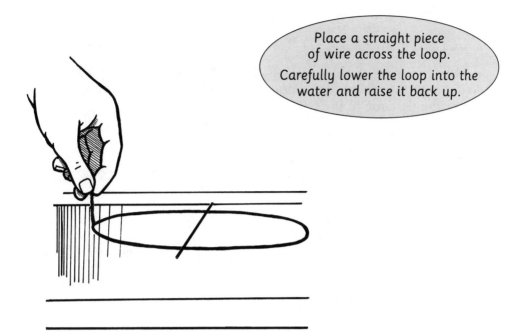

Place a straight piece of wire across the loop.

Carefully lower the loop into the water and raise it back up.

Figure 3-3

5   Use your finger to break the film on one side of the straight wire. What do you see? Try the same experiment without using soapy water. What happens? Why is soapy water necessary for the experiment? (Figure 3-4)

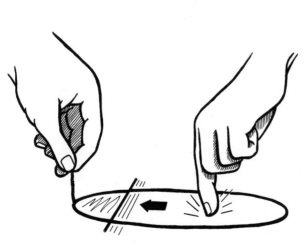

Break the film with your finger.

Figure 3-4

What Happened?

The film of water is stretched equally on each side of the straight wire. When one side of the film is broken, the other side is strong enough to make the wire roll to the edge of the loop.

Guess What?

★ *Dew drops are round because of surface tension.*

★ *When your hair is wet from swimming, it clings together because of surface tension.*

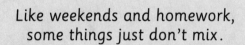

Like weekends and homework,
some things just don't mix.

WEIGHT DEBATE

YOUR CHALLENGE

To compare the weights of different fluids.

DO THIS

1   Pour a small amount of the
    colored water into the glass
    or jar. About 1 inch (2.54 cm)
    deep will do. (Figure 4-1)

Pour water in the jar first.
To make this even more fun,
use your favorite color, or
mix up some colors.

Figure 4-1

YOU NEED

**Water, colored with
food coloring or ink**

**Clear jar or
drinking glass**

Cooking oil

Rubbing alcohol

15

2   Tilt the glass and slowly pour in a layer of
    cooking oil. Pour down the inside edge of
    the glass so the fluids won't
    mix. (Figure 4-2)

Figure 4-2

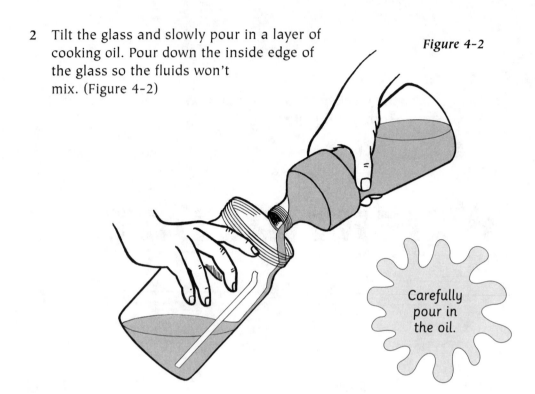

Carefully
pour in
the oil.

3   Next, carefully pour in a layer of rubbing alcohol.
    What do you observe? (Figure 4-3)

Figure 4-3

Know why it's
called rubbing alcohol?

'Cause it's rubbed on the skin to
clean, cool, or disinfect.

Why do you think the fluids did this?
Which fluid is the lightest?
Which is the heaviest?

> Be careful how
> you throw away oil.
> It can pollute the water
> we drink and endanger
> wildlife. Motor oil used in
> cars can be recycled
> again and again.

What Happened?

Fluids will arrange themselves according to their weight. Is water always heavier than oil? Fill a glass about half full with cooking oil and gently place an ice cube on top of the oil. What do you see? What happens when the ice cube melts? Why do you need to shake up some salad dressings before you use them? Which is heavier, oil or vinegar?

Guess What?

★ Water weighs about 8.3 pounds per gallon.

★ A cubic foot of water weighs nearly 62.3 pounds.

★ In the metric system, water has a mass of 1 kilogram per liter.

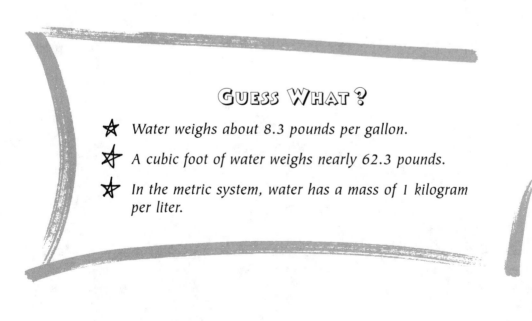

There's no pressure on
you to do this project!

CONTENTS UNDER PRESSURE

YOUR CHALLENGE

To compare the pressure of water at different depths.

DO THIS

1 Use the nail to punch three holes in the side of the jug.
   Make one hole near the bottom, one near the middle, and
   one near the top. Make sure the holes are lined up, one on
   top of the other. (Figure 5-1)

2 Now seal the holes with one long strip of tape.
   (Figure 5-2)

3 Fill the jug with water all of the way to the top. Place the
   jug at the edge of the sink with the holes pointing toward
   the drain.

YOU NEED

**1-gallon plastic milk
jug or 2-liter
plastic bottle**

Nail

**Transparent or
masking tape**

Nail

Water

Kitchen sink

19

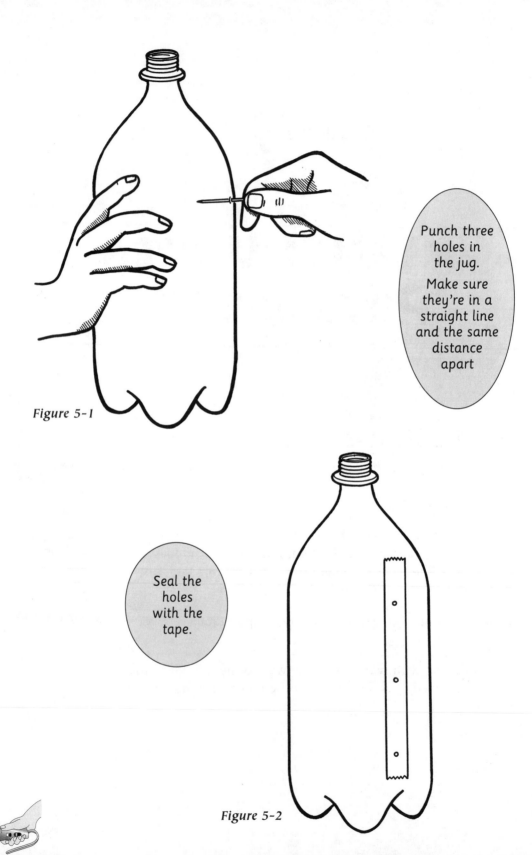

Punch three holes in the jug.

Make sure they're in a straight line and the same distance apart

Figure 5-1

Seal the holes with the tape.

Figure 5-2

20

4   Peel off the tape. What do you see? Which hole has the greatest water
    pressure? (Figure 5-3)

Remove the tape from
the bottom up.

If you don't want your mom to yell
at you, make sure the holes are
pointing toward the sink before
you peel off the tape.

Do you think the shape of a container affects water pressure?
Why or why not?

> Ever get an earache when
> you dive too deep?
>
> If deep-sea divers surface
> too quickly, they can get something
> really scary called decompression
> sickness, or "the bends."

WHAT HAPPENED?

Water weighs a little over 8 pounds per gallon. The more gallons you stack on top of each other, the more pressure you put on the bottom gallon.

What effect does water depth have on divers? Would a diver swimming under an underwater ledge feel less pressure than in open water?

GUESS WHAT?

★ Modern deep-sea divers wearing special suits weighing over 900 pounds can descend to depths of 2,000 feet (600 meters).

★ The pressure of the ocean at its deepest point, about 36,000 feet (11,000 meters) down, is about 6 tons per square inch.

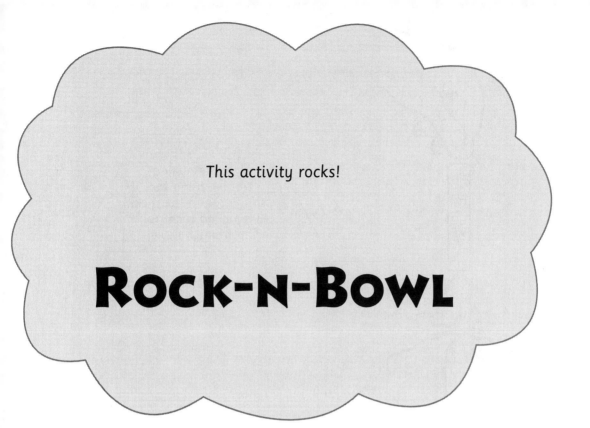

This activity rocks!

ROCK-N-BOWL

YOUR CHALLENGE

To observe the ratio of the density, or compactness, of an object to the density of pure water. This ratio is called the *specific gravity* of the object.

The *specific gravity* of an object is determined by how its density compares with the density of water at 39.2°F (4°C). For example, gold has a specific gravity of 19. This means that any volume of gold is 19 times heavier than the same volume of water.

DO THIS

1 Tie one end of the string around the rock and suspend the rock from the scales. Record the weight. (Figure 6-1)

2 Fill the bowl or bucket with enough water to cover the rock when lowered. (Figure 6-2)

YOU NEED

Rock

String

Spring scales

Water

Large bowl or bucket

Pencil and note pad

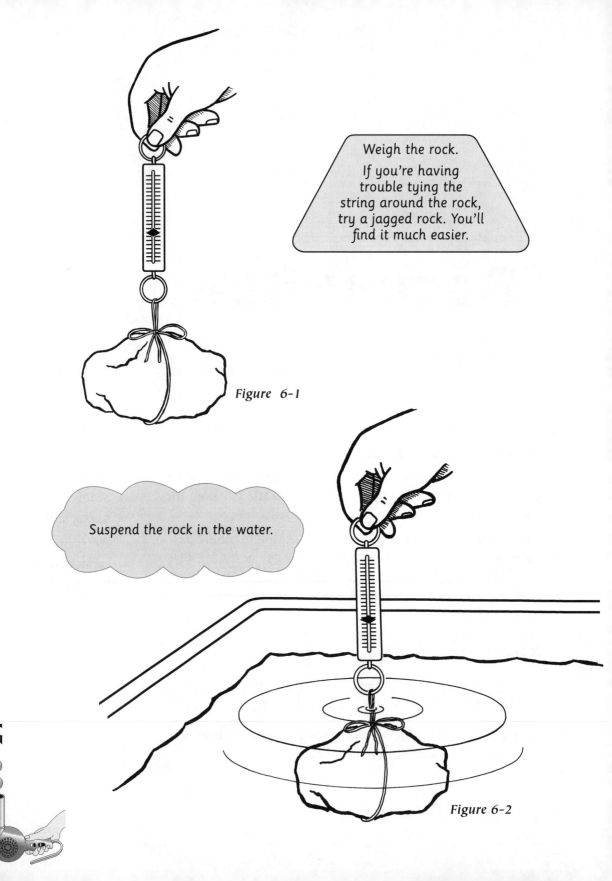

Weigh the rock.

If you're having trouble tying the string around the rock, try a jagged rock. You'll find it much easier.

Figure 6-1

Suspend the rock in the water.

Figure 6-2

3 Now, with the rock still hanging from the scales, lower the rock into the water. The rock should be completely submerged but not touching the bottom of the container.

4 Record the new weight. Compare the two weights.

WHAT HAPPENED?

The specific gravity is the weight of an object in air divided by the amount of weight the object lost when weighed while submerged. For example, if the first weight you recorded was 5 pounds (when the rock was in the air) and the second weight was 3 pounds (when the rock was underwater), you would subtract 3 from 5 to get 2. This means that the rock displaced, or moved, 2 pounds of water. To find the rock's specific gravity, you then divide the rock's weight in air by the weight of the water it displaced, or 5 divided by 2. Therefore, the specific gravity of the rock is 2.5.

How would using saltwater instead of fresh water affect your experiment? Do you think salt water is denser than fresh water? Why or why not? Can you guess why it's easier to float in the ocean than in a pool?

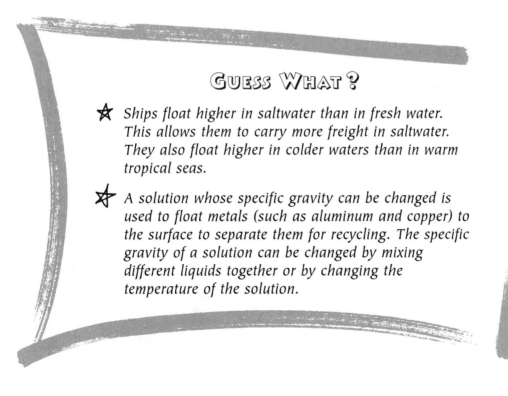

GUESS WHAT?

★ *Ships float higher in saltwater than in fresh water. This allows them to carry more freight in saltwater. They also float higher in colder waters than in warm tropical seas.*

★ *A solution whose specific gravity can be changed is used to float metals (such as aluminum and copper) to the surface to separate them for recycling. The specific gravity of a solution can be changed by mixing different liquids together or by changing the temperature of the solution.*

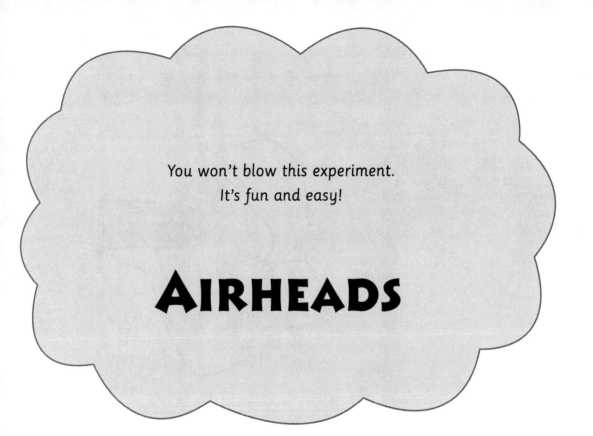

You won't blow this experiment.
It's fun and easy!

AIRHEADS

YOUR CHALLENGE

To trap a volume of air in a confined space.

DO THIS

1 Place the balloon about halfway into the jar and blow up the balloon. What happens? Can you guess why? (Figure 7-1)

2 Release the air from the balloon and try it again. This time, place the straw in the jar. What happens now? (Figure 7-2)

YOU NEED

Clear glass or jar
Balloon
Drinking straw

27

Try to blow up the balloon.
You could blow until you're blue
in the face. It's impossible!

Figure 7-1

Figure 7-2

Now, add a straw and
blow up the balloon.

Know why this is easier?
Because an empty jar is
not really empty.
It's full of air!

WHAT HAPPENED?

The trapped air exerted a pressure on the balloon equal to the pressure of the balloon on the air. Why did the balloon expand outside of the jar? What was the purpose of the straw?

GUESS WHAT?

★ *Some shock absorbers on cars use trapped air to cushion the ride.*

★ *It is impossible to remove all of the air in an airtight container.*

Just chill and have fun!

ICE CAP

YOUR CHALLENGE

To observe the pressure of the weight of the atmosphere.

DO THIS

 1 Using the funnel, fill the plastic bottle about one-third full with hot tap water. (Figure 8-1)

2 Wait a couple of minutes, and then screw on the cap. (Figure 8-2)

3 Lay the bottle in the sink and pour ice and cold water over it. What happens to the bottle? (Figure 8-3)

YOU NEED

Sink

Funnel

Plastic soda bottle with cap

Hot tap water

Cold water

Ice

31

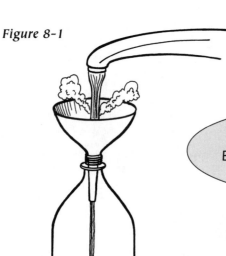

Figure 8-1

Fill the bottle with hot water.
Be careful! Tap water can scald
if it's too hot.

Wait a little, then screw on the lid.
Don't have an empty bottle? Here's a
good excuse to finish off that soda
in the fridge!

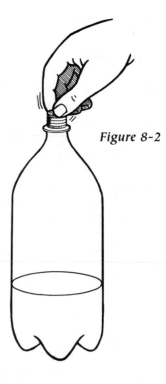

Figure 8-2

WHAT HAPPENED?

As warm air inside the bottle cools, it exerts less pressure on the sides of the bottle. The pressure of the air outside crushes the bottle inward.

Atmospheric pressure pushes in all directions. However, the pressure gradually decreases the higher you go. At sea level the pressure is almost 15 pounds per square inch (1000 millibars).

Chill the bottle.

Did you know that hot air weighs less than cold air? Airplanes take off more easily in cold air than hot air, but balloons rise using hot air. Can you guess why?

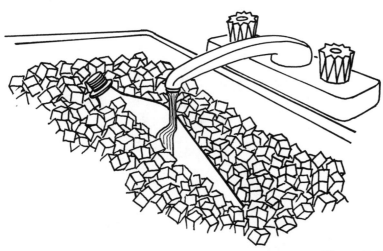

Figure 8-3

GUESS WHAT ?

★ Our bodies exert as much pressure out as the pressure of the atmosphere pushes in; otherwise, we would be crushed.

★ Without the weight of air, you couldn't suck liquid, such as lemonade or soda pop, through a straw.

COMPRESSION SESSION

YOUR CHALLENGE

To observe that air can be squeezed together, or compressed.

DO THIS

1 Fill the bowl with water.

2 Hold the glass upside down and slowly push it down to the bottom of the bowl. Notice any escaping air bubbles. Were there any? (Figures 9-1 and 9-2)

3 Look at the water level inside the glass. What happened to the air inside the glass? Does it take up less space? Why do you think the water level changed?

YOU NEED

Large bowl

Water

Clear drinking glass

35

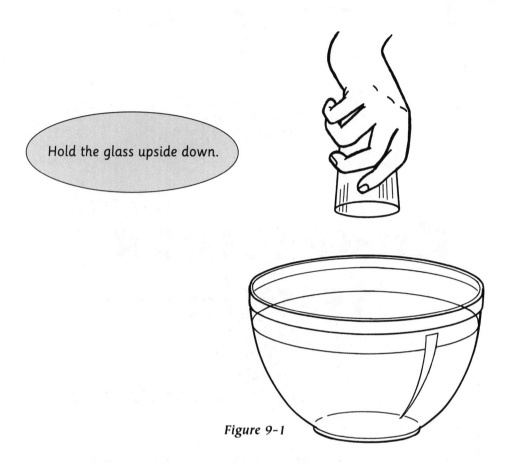

Hold the glass upside down.

Figure 9-1

Figure 9-2

Push the glass to the bottom of the bowl.

Did you know that the air in our atmosphere is made mostly of nitrogen and oxygen?

What Happened?

Small particles of air, called *air molecules*, can be squeezed together, or compressed. Compressed air has many uses. For example, compressed air is used to inflate a basketball or a bicycle tire. It can also be used to power machines and even to stop trains.

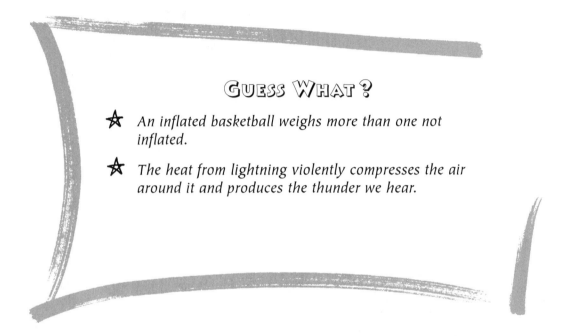

Guess What?

★ An inflated basketball weighs more than one not inflated.

★ The heat from lightning violently compresses the air around it and produces the thunder we hear.

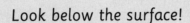

Look below the surface!

DIVING MEDICINE DROPPER

YOUR CHALLENGE

To compare the differences between compressing air and water.

DO THIS

1 Fill the plastic bottle completely with water. (Figure 10-1)

Plastic soda bottle with cap

Water

Medicine dropper

Fill the bottle with water.
Save the cap.
You're going to need it!

Figure 10-1

39

2   Fill the medicine dropper with water so that the rubber end will float just a little above the surface.

3   Drop the medicine dropper, rubber end up, into the bottle. Screw the cap back on the bottle. (Figure 10-2)

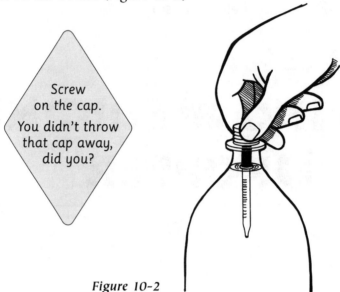

Screw
on the cap.
You didn't throw
that cap away,
did you?

Figure 10-2

4   Squeeze the bottle. What happens to the medicine dropper? Why do you think this happens? (Figure 10-3)

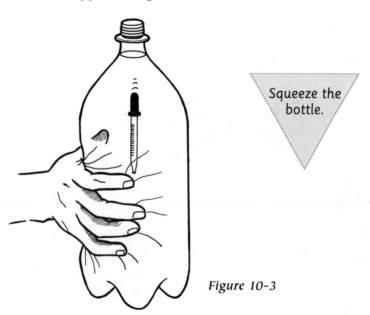

Squeeze the
bottle.

Figure 10-3

5 Release your grip on the bottle. What do you observe? Which compresses easier, water or air? (Figure 10-4)

Release the pressure.

Figure 10-4

Did you know that the word scuba comes from Self-Contained Underwater Breathing Apparatus?

WHAT HAPPENED?

With gentle squeezing, the medicine dropper can be made to float at any level in the bottle. Air is easier to compress than water. When air is compressed, it takes up less space. With less space, an object will float lower in the water.

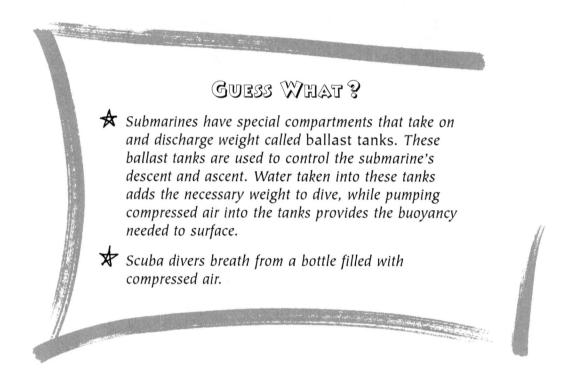

GUESS WHAT?

⭐ *Submarines have special compartments that take on and discharge weight called* ballast tanks. *These ballast tanks are used to control the submarine's descent and ascent. Water taken into these tanks adds the necessary weight to dive, while pumping compressed air into the tanks provides the buoyancy needed to surface.*

⭐ *Scuba divers breath from a bottle filled with compressed air.*

Have a blast!

BOTTLE ROCKET

YOUR CHALLENGE

To build and fly a model rocket.

DO THIS

 1 Use the scissors to carefully cut and shape four cardboard fins for the rocket. Make sure the fins will extend past the opening in the bottle about 4 inches (10 cm). (Figure 11-1)

2 Shape the fins to fit the bottle. (Figure 11-2)

3 Fasten the fins to the bottle with masking tape. (Figure 11-3)

 4 Place the cork on a piece of scrap wood and use the hammer and nail to make a small hole in the center of the cork. This hole should make a snug fit for the inflating needle of the pump. (Figure 11-4)

YOU NEED

2-liter plastic soda bottle

Thin cardboard

Scissors

Masking tape

Cork that fits bottle

Hammer

Nail

Funnel

Water

Bicycle hand pump and inflating needle (used for basketballs)

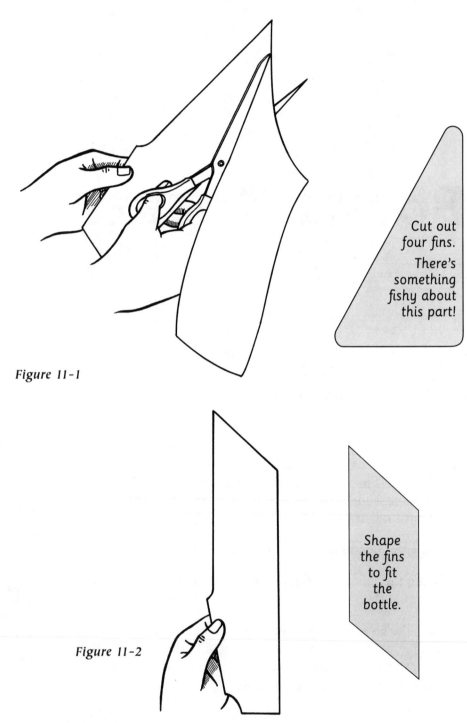

Cut out
four fins.
There's
something
fishy about
this part!

Figure 11-1

Shape
the fins
to fit
the
bottle.

Figure 11-2

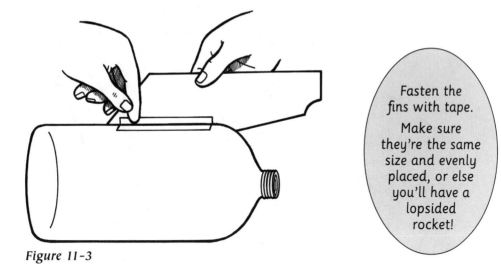

Figure 11-3

Fasten the fins with tape.

Make sure they're the same size and evenly placed, or else you'll have a lopsided rocket!

5   Using the funnel, fill the bottle about one-fourth full of water and push the cork into the top of the bottle. (Figure 11-5)

6   Take the rocket outside to a large, open area.

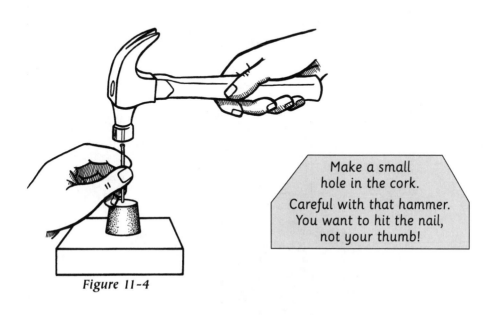

Figure 11-4

Make a small hole in the cork.

Careful with that hammer. You want to hit the nail, not your thumb!

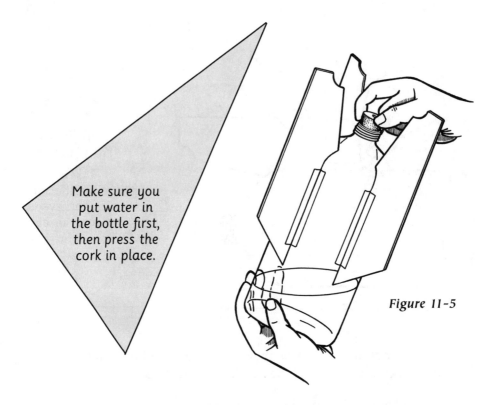

Make sure you put water in the bottle first, then press the cork in place.

Figure 11-5

7   Attach the needle to the tire pump and insert the needle through the hole in the cork. (Figure 11-6)

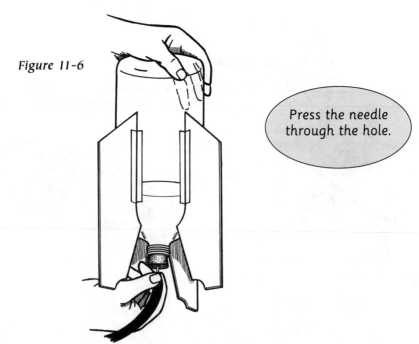

Figure 11-6

Press the needle through the hole.

8 Stand the rocket on its fins and stretch out the hose to the pump so that you can stand clear of the rocket. (Figure 11-7)

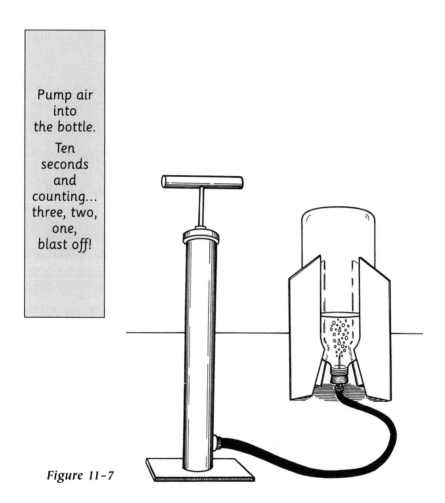

Pump air into the bottle.

Ten seconds and counting... three, two, one, blast off!

Figure 11-7

 9 Start pumping vigorously. What happens? Can you guess why?

Experiment with different amounts of water in the rocket. Record the results to find which one produced the best performance.

What Happened?

For every action, there is an equal but opposite reaction. If you push or pull anything, it pushes or pulls right back. How does this compare to the operation of your rocket? What happens to the end of a garden hose if the water is suddenly turned on?

In space, where there is no air to push against, the rocket is propelled by the gases from the engine pushing the rocket forward and the gases back. For an instant at the top of its flight, the rocket became weightless.

The terms *weight* and *mass* are often confused. *Weight* is the measurement of the force of gravity. *Mass* is the amount of matter in an object. In space, astronauts have the same mass as they do on earth, but most of the time they have little or no weight.

Two other terms that are often mixed up are *speed* and *velocity*. *Speed* tells us the rate at which an object travels. *Velocity* tells us the speed *and the direction* of an object. For example, if one student is riding a bicycle north at 10 miles (16 km) per hour and another is riding south at 10 miles (16 km) per hour, they both have the same speed, but not the same velocity. To have the same velocity they must be traveling in the same direction.

Acceleration is the increase of speed of an object during a certain period of time. A car accelerates when, for instance, a traffic light turns from red to green and the driver puts his or her foot on the gas pedal. However, the car is not accelerating when it is traveling down a highway at a steady speed.

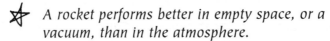

Guess What?

✶ A rocket performs better in empty space, or a vacuum, than in the atmosphere.

✶ The two huge booster rockets that lift the space shuttle to the upper atmosphere are over 149 feet (45 meters) tall and about 12 feet (4 meters) wide.

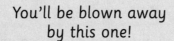

You'll be blown away
by this one!

FLOATING PING-PONG BALL

YOUR CHALLENGE

To observe the effect of
a column of air on a
round object.

DO THIS

1 Turn the hair dryer on
   high without heat.

2 Aim the nozzle straight
   up and place the
   Ping-Pong ball in the
   center of the column of
   air. Release the ball.
   What does the
   ball do? Why do you
   think the ball behaves
   this way?
   (Figure 12-1)

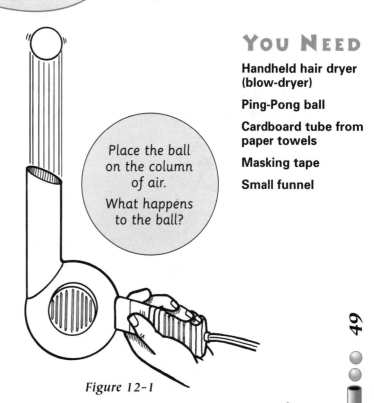

Place the ball
on the column
of air.
What happens
to the ball?

Figure 12-1

YOU NEED

**Handheld hair dryer
(blow-dryer)**

Ping-Pong ball

**Cardboard tube from
paper towels**

Masking tape

Small funnel

3   Move the ball slightly to one side of the column of air and release it. Now what does the ball do?

4   Tilt the hair dryer (the column of air) slightly to one side. What happens to the ball now?

5   Turn the dryer off and tape a cardboard tube to the opening of the dryer. (Figure 12-2)

Tape a tube to the dryer.
Don't forget to take the tube off
next time you dry your hair!

Figure 12-2

6   Place the small end of the funnel in the end of the tube. Hold or tape the funnel to the tube. (Figure 12-3)

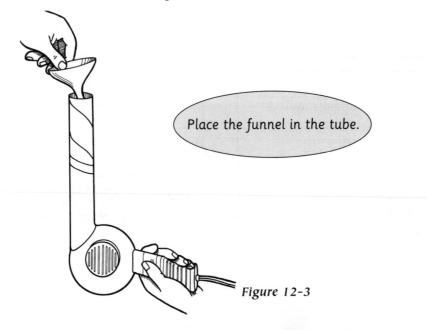

Place the funnel in the tube.

Figure 12-3

7 Turn on the dryer and place the Ping-Pong ball in the funnel. Now what does the ball do? (Figure 12-4)

Turn on the dryer.

Figure 12-4

8 Point the dryer down. Now what happens to the ball?

WHAT HAPPENED?

The faster air moves, the more the pressure of the air drops. Can you explain why the ball behaves as it does?

GUESS WHAT?

⭐ Rising columns of air in a cumulus cloud can exceed 33 feet (110 meters) per second. (Cumulus clouds are the puffy kind, sometimes called "fair-weather" clouds.)

⭐ The very low air pressure in a tornado can cause trees and telephone poles to expand. This happens because the normal air trapped in tiny cells inside the wood is unable to adjust to the rapidly reduced air pressure outside.

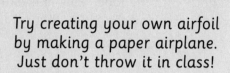

Try creating your own airfoil
by making a paper airplane.
Just don't throw it in class!

FOILED AGAIN!

YOUR CHALLENGE

To observe how moving air creates lift.

DO THIS

1 Fold the paper in half so that it is about 5 inches
   (12.5 cm) long. (Figure 13-1)

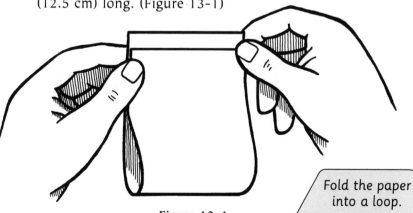

Figure 13-1

Fold the paper
into a loop.

Be sure the loop
has an edge.

YOU NEED

**Sheet of paper about
3 inches (8 cm)
wide and 10 inches
(25 cm) long**

Cardboard

Transparent tape

Electric fan

2   Hold the ends together and tape the paper to the cardboard. Before
    sticking the tape, however, move the end of the top of the paper so
    that the paper opens slightly into the loop. This makes the bottom of
    the loop flat with the cardboard, and the top half curved.
    (Figure 13-2)

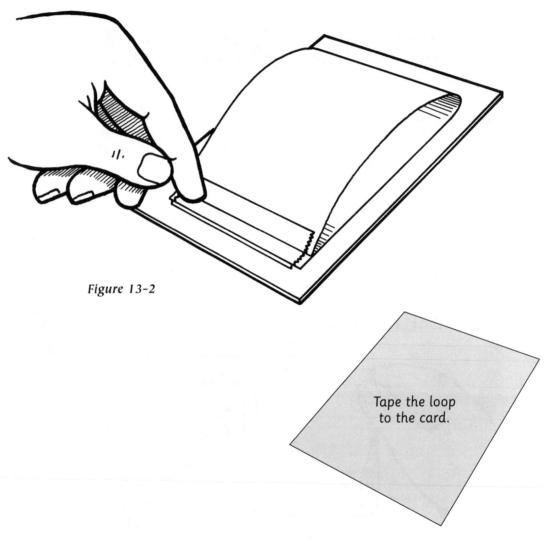

Figure 13-2

Tape the loop
to the card.

3   You have just made an *airfoil*. Stick the airfoil to the cardboard. (Figure 13-3)

Figure 13-3

Hold the loop in the breeze of a fan. Hold on tight!

4   Hold the cardboard in front of the fan with the tape ends pointing toward the fan. Turn the fan on HIGH.

What do you observe? Why do you think the airfoil does this? What is the difference between the speed of the air above and below the paper?

WHAT HAPPENED?

The speed of the moving air over or around a curved surface tends to reduce the pressure of the air above that surface.

Where else could you find airfoils? Would the wings of a plane be airfoils? How about the blades of a propeller? Can cars and boats have airfoils, too?

GUESS WHAT?

★ A jumbo jet must reach speeds of about 180 miles per hour (290 km per hour) before the wings create enough lift for it to fly.

★ The blades of a helicopter create lift just like the wings of an airplane. The whop-whop sound you hear is caused by the tips of the blades breaking the sound barrier.

This experiment's all wet!

WATER SPRAYER

YOUR CHALLENGE

To observe how reduced air pressure can lift a liquid.

DO THIS

 1   Have an adult carefully cut a slit crossways in the straw about 2 inches (5 cm) from one end. Be sure not to cut all the way through. (Figure 14-1)

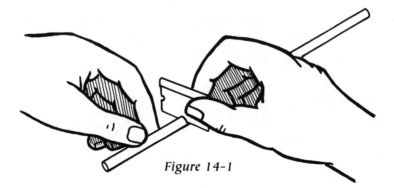

Figure 14-1

Have an adult cut a slit in the straw.

57

2 Flatten the longer part of the straw and bend it at the cut. (Figures 14-2 and 14-3)

Flatten part of the straw.

If you cut too much of the straw, try again.

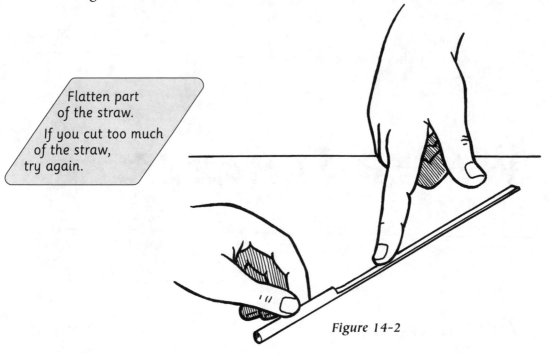

Figure 14-2

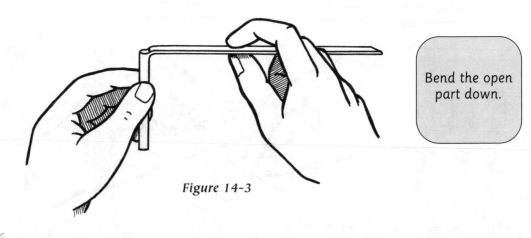

Figure 14-3

Bend the open part down.

3 Put the short end in the glass of water so that the bend is just above the surface of the water and at the far edge of the glass. Blow hard through the straw. (Figure 14-4)

Blow through the straw.

Figure 14-4

What do you see? Can you explain what happens? How does atmospheric pressure affect your experiment? How does this experiment compare to a perfume sprayer?

WHAT HAPPENED?

The faster any liquid moves, the lower the pressure. Airbrushes, which are used to spray paint, operate on low air-pressure systems. Carburetors in gasoline engines have a narrow chamber that speeds up an airflow, drawing in a fine spray of fuel. The fuel is then fed to the cylinders, where it is ignited.

GUESS WHAT?

★ *Many drive-in windows at banks use a vacuum system to handle customers' transactions.*

★ *The sprayer at a kitchen sink works on the same principle as your experiment.*

PART TWO
SOLID MECHANICS

Solid mechanics involves dynamics and statics. *Dynamic* means energetic, vigorous, or forceful. In physics, *dynamics* deals with the forces affecting objects or bodies in motion. The energy of a moving object is called *kinetic energy*. A falling rock, for instance, has kinetic energy.

Statics deals with bodies, masses, or forces at rest. Bodies at rest have *potential energy*. For example, the same rock sitting on the edge of a cliff has potential energy, not kinetic energy.

In the experiments that follow, we will build simple machines to demonstrate some principles of solid mechanics. Normally, we think of a machine as something with a motor that performs work. But a machine is really any device that does work. It doesn't have to have a motor.

A machine produces a force and controls the direction and speed of the force, but it cannot produce energy on its own. It can never do more work than the amount of energy put into it because of the *friction*, or rubbing together, of its parts. The *lever* is probably the most efficient simple machine. The work it puts out nearly equals the energy put in because the energy lost to friction is very small.

Are you inclined
to do this project?

RAMP ART

YOUR CHALLENGE

To observe how one of the simplest machines reduce works.

DO THIS

1   Stack the books about 6 inches (15 cm) high and make two ramps with the rulers. (Figure 15-1)

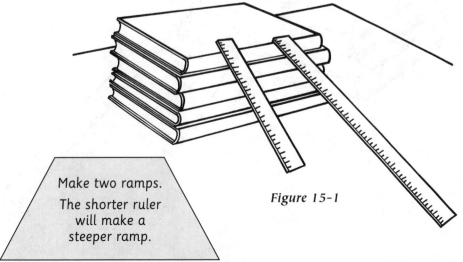

Make two ramps.

The shorter ruler will make a steeper ramp.

Figure 15-1

YOU NEED

Four or five thick hardback books

12-inch (15-cm) ruler

36-inch (1 meter) ruler

Hammer

Three nails

Rubber band

Wooden block about 2 x 4 x 8 inches (5 x 10 x 20 cm)

Metal file

2 Drive one of the nails into the end of the block of wood or have an adult do it. Attach the rubber band to the nail and pull the block of wood up the ramps. Notice how far the rubber band stretches. How does the steepness of the ramp affect the rubber band? (Figure 15-2)

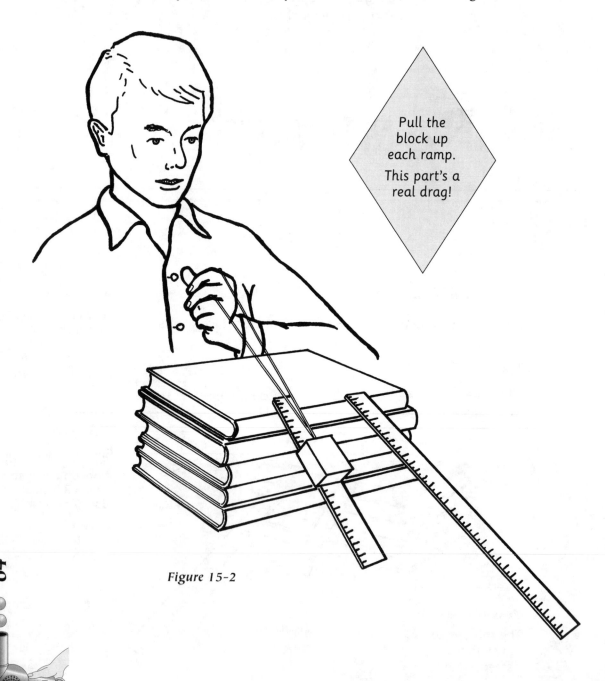

Pull the block up each ramp.

This part's a real drag!

Figure 15-2

3   Carefully file the point of one of the nails so it is blunt. Now hammer it into the block of wood. (Figures 15-3 and 15-4)

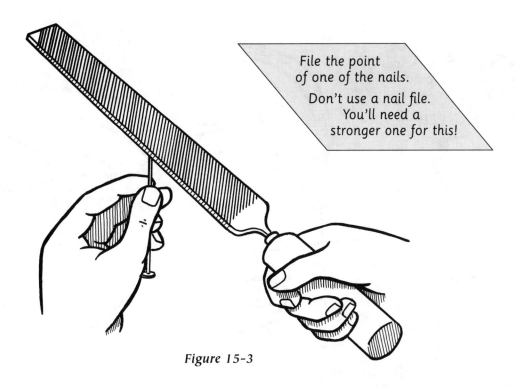

File the point
of one of the nails.

Don't use a nail file.
You'll need a
stronger one for this!

Figure 15-3

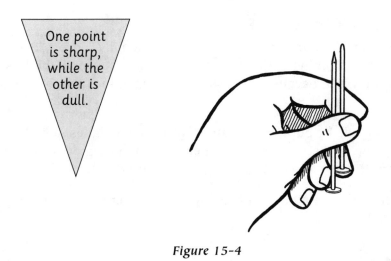

One point
is sharp,
while the
other is
dull.

Figure 15-4

4 Hammer in the nail with the point. Notice how much more easily the pointed nail goes in. Can you explain why? (Figure 15-5)

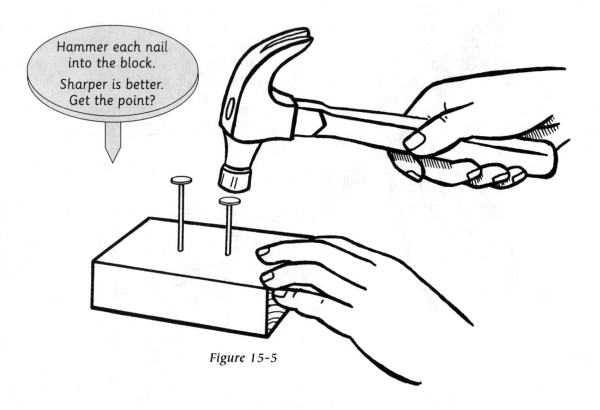

Hammer each nail into the block.

Sharper is better. Get the point?

Figure 15-5

What Happened?

A *ramp* is a machine that makes it easier to gradually move a load to a height instead of lifting it straight up. The longer the slope, the less effort required. However, because the distance is longer, the amount of work is the same.

A *wedge* is made by putting two ramps, or inclined planes, together. The sloping point on the nail makes the penetration more gradual so it takes less effort. Where can you find wedges being used?

Why are roads up a mountain twisting and curving instead of going straight to the top?

GUESS WHAT?

★ *Water freezing in the cracks of boulders often forms ice wedges that break off huge pieces of rock.*

★ *Axes, chisels, and needles are some examples of wedges.*

★ *Nails are sized by the "penny"—a 1-penny nail, a 2-penny nail, a 3-penny nail, and so on. This is because nails were once sold by the pence, or penny, in England.*

You don't have to have a screw loose
to do this one!

As the Screw Turns

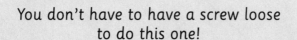

Your Challenge

To investigate another form of
inclined plane.

Do This

 1 Screw the screw into the
wooden block. Notice how
the threads of the screw
cut into the wood.
(Figure 16-1)

> Turn the screw
> into the block.
>
> Here's a tip: Ask an adult to make
> a shallow starter hole with a nail.
> It will make the screw
> go in more easily.

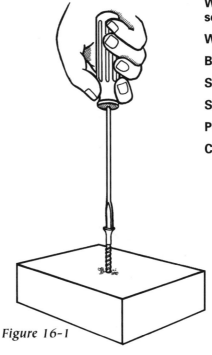

Figure 16-1

You Need

**Woodscrew
screwdriver**

Woodscrew

Block of wood

Sheet of paper

Scissors

Pencil

Colored marker

2 With a pair of scissors, cut a right triangle out of the sheet of paper that is a little shorter than the pencil to make a ramp shape. (Figure 16-2)

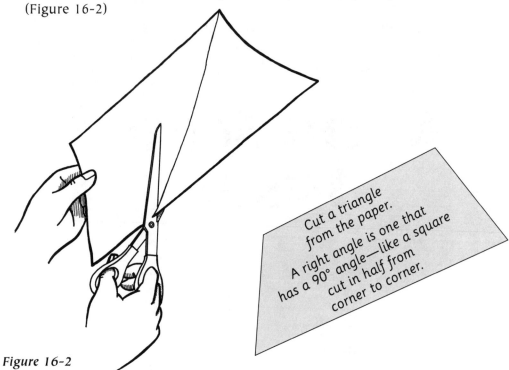

Cut a triangle from the paper.
A right angle is one that has a 90° angle—like a square cut in half from corner to corner.

Figure 16-2

3 Use a marker to make a colored line along the slanting edge of the paper. This will make the spiral stand out. (Figure 16-3)

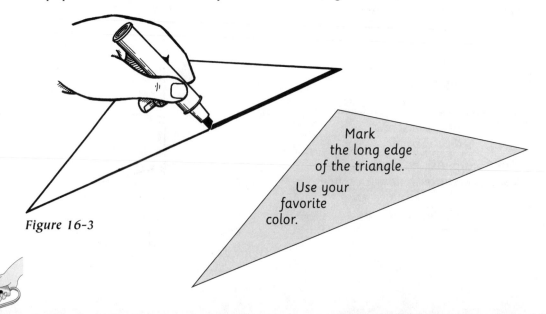

Mark the long edge of the triangle.

Use your favorite color.

Figure 16-3

4   Roll the paper on the pencil from the short side of the triangle to the point. Keep the bottom, or baseline, of the triangle even as it rolls. (Figures 16-4 and 16-5)

Roll the paper onto the pencil.

Now you can see how the threads that wrap around a screw are really an inclined plane!

Figure 16-4

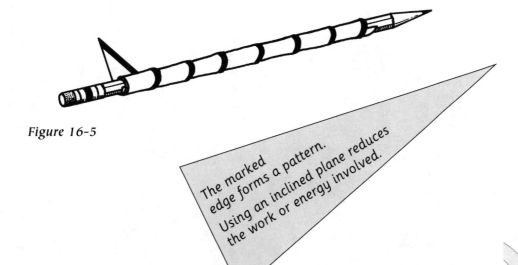

Figure 16-5

The marked edge forms a pattern. Using an inclined plane reduces the work or energy involved.

What pattern is formed by the marked edge? Which requires more force, driving a nail or turning a screw? Which is the strongest? Why do you think this happens? How do the jaws of a vise grip and hold objects?

WHAT HAPPENED?

A screw can be thought of as an inclined plane wrapped around a rod.

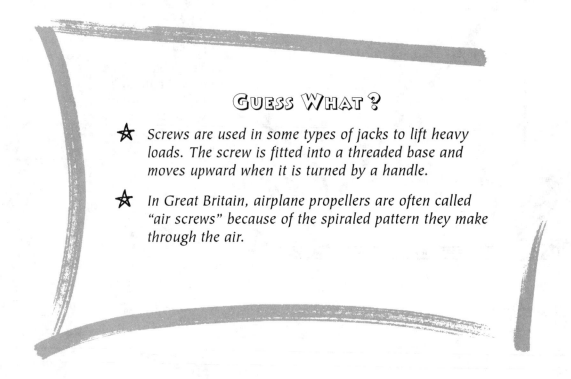

GUESS WHAT?

★ Screws are used in some types of jacks to lift heavy loads. The screw is fitted into a threaded base and moves upward when it is turned by a handle.

★ In Great Britain, airplane propellers are often called "air screws" because of the spiraled pattern they make through the air.

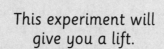

This experiment will give you a lift.

CLEVER LEVERS

YOUR CHALLENGE

To investigate different types of levers.

DO THIS

1 Stand at one end of the table and try lifting the legs off the floor. Notice how much force is required. (Figure 17-1)

Lift the table. If this is too much like work, go lie down and rest when you're done!

YOU NEED

Desk or table

Two wooden boards, 4 or 5 inches (10 or 12 cm) wide and about the same height as the table

Pair of pliers

Baseball bat or tennis racket

Figure 17-1

73

2 Place one of the boards on its end next to the table and the other
   board on top.

3 With one end of the board under the edge of the table, press down on
   the other end. Compare the amount of force you are now using. Can
   you explain the difference? Where is the fulcrum (point of support)?
   (Figure 17-2)

Press down on the
table to lift the board.

Does this make you think of Archimedes?
(Don't know who Archimedes was?
Check the encyclopedia!)

Figure 17-2

4 Move the board standing on end away from the table a little and try it
   again. Now how much force are you using? (Figure 17-3)

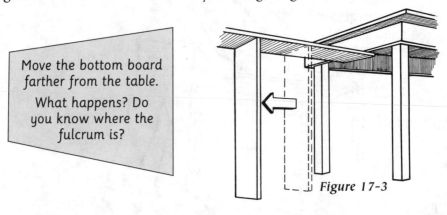

Move the bottom board
farther from the table.

What happens? Do
you know where the
fulcrum is?

Figure 17-3

5   In a wide-open space outside, swing the tip of the bat as though you
    were hitting a ball. (Figure 17-4)

Move the tip of the
bat in an arc.

So it's not just how
hard you swing, it's
where the bat hits the
ball that makes a
home run!

Where is
the fulcrum
on a swing?

Figure 17-4

How far does the tip of the bat move compared to the handle? Where is
the force applied and where is the fulcrum?

What Happened?

A *lever* is a stiff bar that turns around a pivotal point called a *fulcrum*. The
advantage of using a lever comes from the short distance between the
load and the fulcrum compared to the long distance between the fulcrum
and the point where the force is applied. A *force* can be thought of as any
push or pull that acts in a certain direction.

75

There are three basic types of levers. In the first type, the fulcrum is between the point where the force is applied and the load. You used this type of lever when you lifted the table. In the second type of lever, the load is located between the fulcrum and the point where the force is applied. A wheelbarrow uses this principle. The axis of the wheel is the fulcrum and the force is applied to the handles to lift the load. These first two types of levers are called *force-multipliers*.

In the third type of lever, the force doing the work is applied between the load and the fulcrum. This type is called a *motion-multiplier*. It can multiply speed as well as distance. Brooms and fishing rods use this principle.

How many other places can you think of where levers are used?

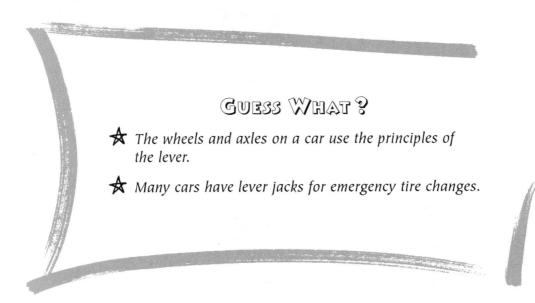

Guess What?

⭐ *The wheels and axles on a car use the principles of the lever.*

⭐ *Many cars have lever jacks for emergency tire changes.*

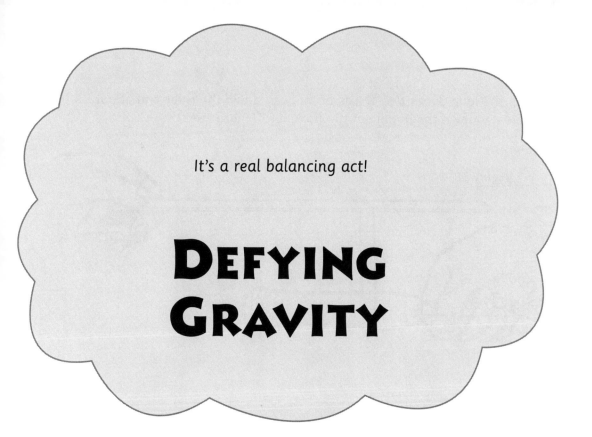

It's a real balancing act!

DEFYING GRAVITY

YOUR CHALLENGE

To observe how the center of gravity affects balance. The *center of gravity* is the one point in an object, or an assembly of objects, where the weight is evenly balanced.

DO THIS

1   Tie the string to form a loop about 4 or 5 inches (10 or 12 cm) long. (Figure 18-1)

Make a loop with the string.

Make sure the knot won't slip.

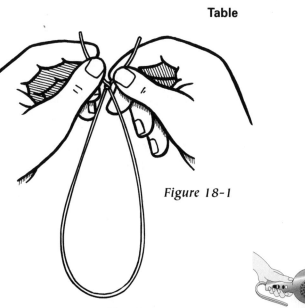

Figure 18-1

77

2 Slip the loop about two-thirds of the way down the ruler and about halfway down the handle of the hammer. (Figure 18-2)

Figure 18-2

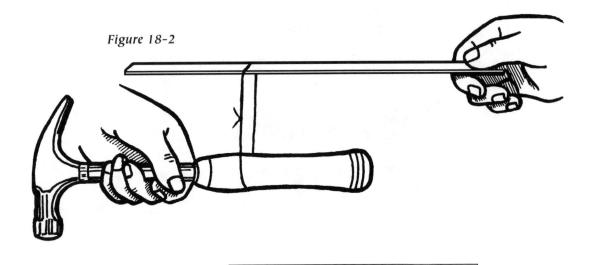

> Slide the loop over the hammer and the ruler.
>
> You might need to adjust the hammer to get just the right spot.

3 Holding the hammer in place, put the tip of the ruler on the edge of a table. The head of the hammer should be slightly under the table and the end of the handle about one-third of the way down the ruler.

4 Release the hammer. Do the ruler and hammer stay in place? Why do you think this happens? (Figure 18-3)

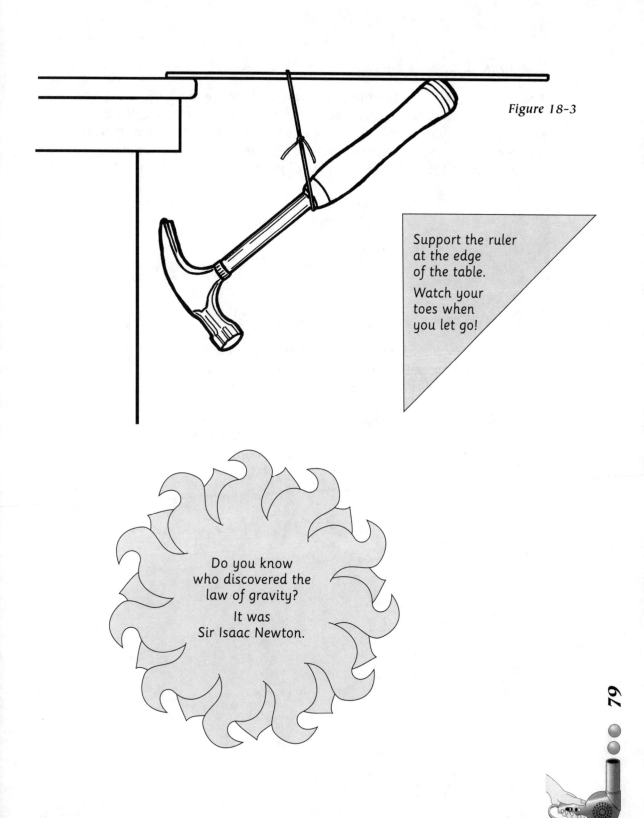

Figure 18-3

Support the ruler
at the edge
of the table.

Watch your
toes when
you let go!

Do you know
who discovered the
law of gravity?

It was
Sir Isaac Newton.

79

WHAT HAPPENED?

The *center of gravity* is that one point where all of the mass of an object seems to exist. Because gravity acts on mass to create weight, this one point is where gravity acts on the object to create the object's weight. If an upward force equal to the object's weight is exerted at this point, it will be balanced. Where is the center of gravity in this experiment?

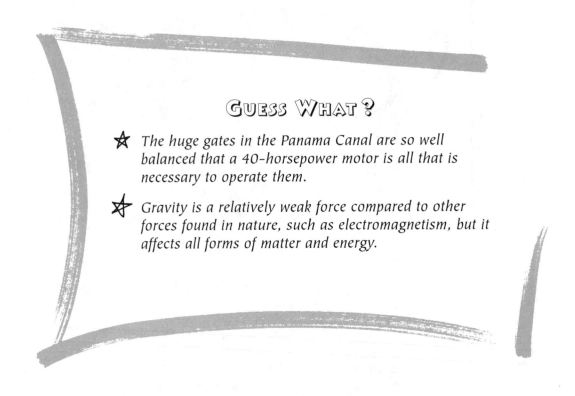

GUESS WHAT?

★ The huge gates in the Panama Canal are so well balanced that a 40-horsepower motor is all that is necessary to operate them.

★ Gravity is a relatively weak force compared to other forces found in nature, such as electromagnetism, but it affects all forms of matter and energy.

This project will show
you the ropes!

PULLEY EASE

YOUR CHALLENGE

To observe and compare the lifting power of a fixed and movable pulley.

DO THIS

 1 Use the pliers to cut the bottom section from a coat hanger.

2 Slide the spool onto the center of the wire.

3 Bend the free ends of the wire up and loop them together. This will be your pulley. Hang the pulley from a support, such as a coat hook or a doorknob. (Fig. 19-1)

4 Tie the string around the books and thread the end over the top of the spool-pulley. This is called a *fixed* pulley.

5 Pull down on the end of the string. Notice how much force it takes to lift the books. (Figure 19-2)

Wire-cutter pliers

Wire clothes hanger

Empty thread spool

String

Two or three books for weights

81

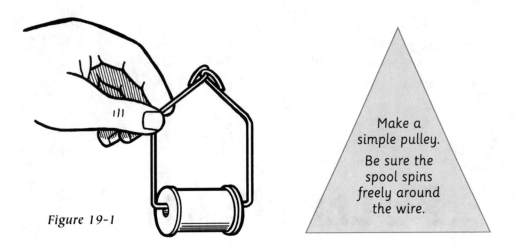

Figure 19-1

Make a simple pulley.

Be sure the spool spins freely around the wire.

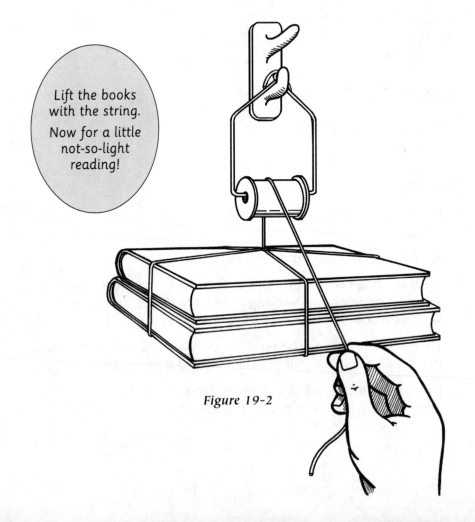

Lift the books with the string.

Now for a little not-so-light reading!

Figure 19-2

6   Remove the string and the pulley from its support.

7   Use a short length of string to fasten the pulley, upside down, to the books. The twisted-wire part of the pulley should be attached to the string that bundles the books. This pulley is considered *movable*.

8   Tie one end of a longer string to the support and run it around the pulley.

9   Pull up on the string to lift the books. How much force is needed now? Why do you think this amount of force is now required? (Figure 19-3)

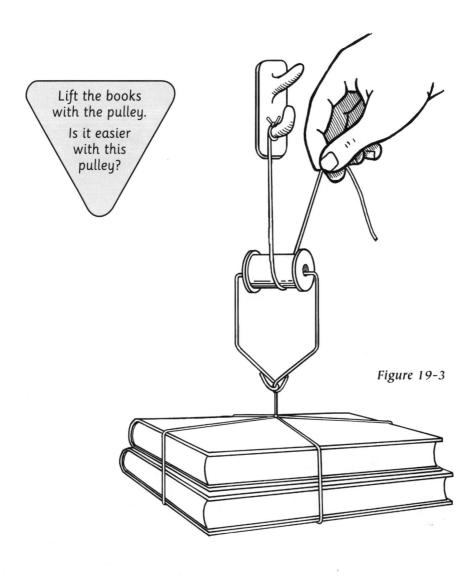

Lift the books with the pulley.

Is it easier with this pulley?

Figure 19-3

What Happened?

With only one pulley, the force to lift an object is the same as the weight of the object. The only advantage with one pulley is that you can pull from a different direction and add your weight to the force of the pull. Using a movable pulley reduces the force required by half. What would happen if you used two pulleys?

Where can you find pulleys around the home? Are window drapes operated by pulleys? How is the flag raised at your school?

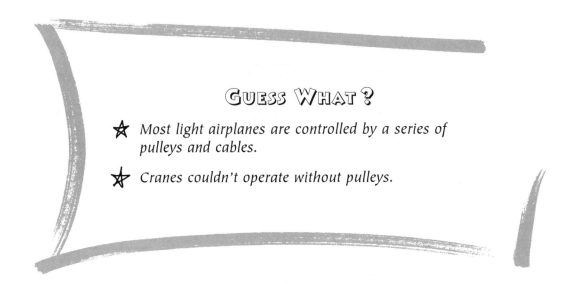

Guess What?

★ Most light airplanes are controlled by a series of pulleys and cables.

★ Cranes couldn't operate without pulleys.

Don't cause any static.
Just do the project!

FRICTION PREDICTION

YOUR CHALLENGE

To compare the forces necessary to overcome *static friction*, the force to move an object at rest, and *moving friction*, the force required to keep an object moving.

DO THIS

1 Connect the rubber bands together and attach one end to the shoe box. (To attach, you can poke the rubber band through the box and put a short pencil through the loop, as shown in the illustration.) Place a book or something inside for added weight. (Figures 20-1 and 20-2)

2 Place the box on a smooth floor or table and pull on the other end of the rubber bands. Notice how far they stretch before the box starts to move. (Figure 20-3)

YOU NEED

Several rubber bands

Shoe box with something inside for weight, such as a book

Four pencils

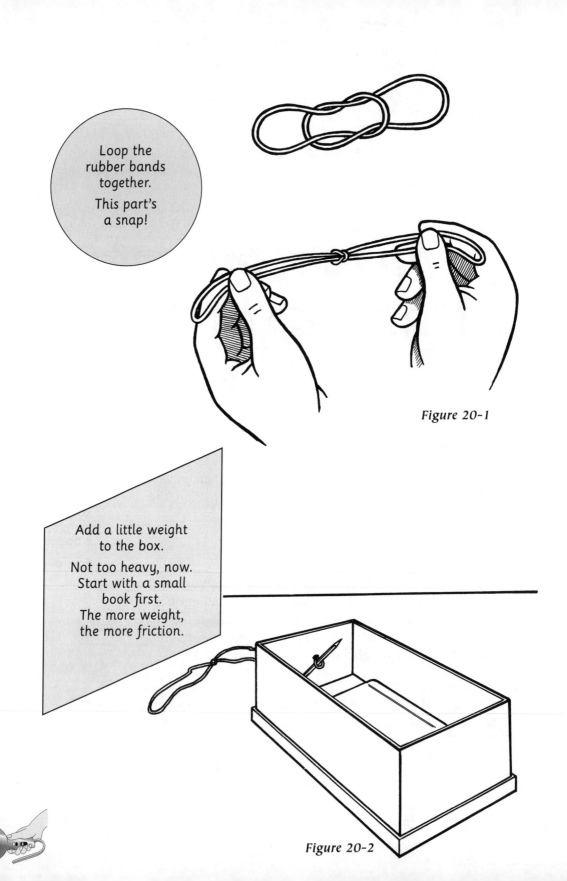

Loop the rubber bands together.

This part's a snap!

Figure 20-1

Add a little weight to the box.

Not too heavy, now. Start with a small book first. The more weight, the more friction.

Figure 20-2

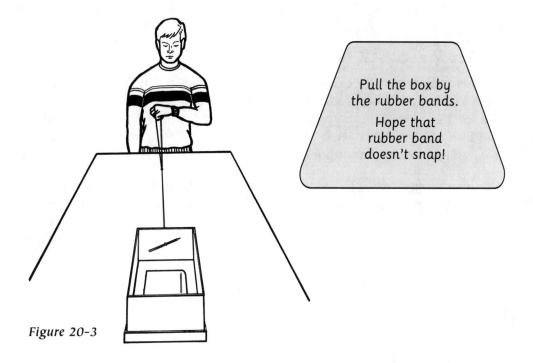

Figure 20-3

3   After the box is moving at a steady pace, notice how far the rubber bands are stretched now. How do they compare?

4   Experiment by placing pencils under the box to act as rollers or by putting the box on carpet. (Figure 20-4)

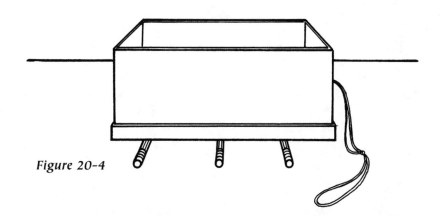

Figure 20-4

WHAT HAPPENED?

Friction is a force that occurs when one object moves while it is in contact with another. Friction produces heat. When an automobile begins moving, it uses a transmission to go from the lower gears to the high gear it uses to travel at highway speeds. This means that it takes more force to start something moving than it does to keep it moving.

Can you think of ways to reduce the friction? Rub your hands together. What happens? Can air cause friction? How does this affect the design of cars?

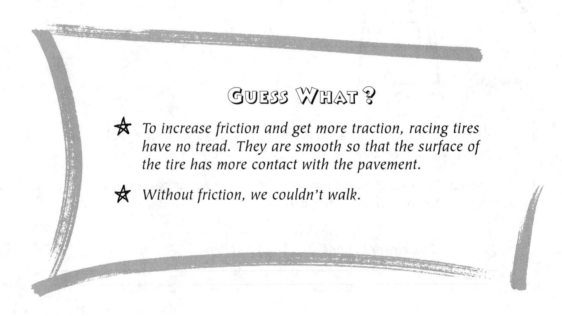

GUESS WHAT?

★ *To increase friction and get more traction, racing tires have no tread. They are smooth so that the surface of the tire has more contact with the pavement.*

★ *Without friction, we couldn't walk.*

Have a ball
with this one!

GET YOUR
BEARINGS
STRAIGHT

YOUR CHALLENGE

To observe the effect of ball bearings on friction.

DO THIS

1   Place one can on top of the
    other. Place books on top of
    the cans and try turning the
    top can. Notice how much
    force is used to move the
    can. Remove the top can.
    (Figure 21-1)

YOU NEED

Marbles

**Two metal cans, the
same size, with
grooves around the
top (paint cans
work well)**

**Stack of hardback
books**

Try turning the
top can.

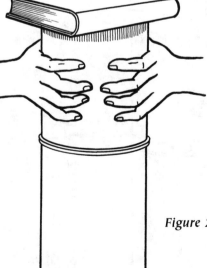

Figure 21-1

2   Now place marbles in the groove in one of the cans. Invert the other
    can and place it on top so that the groove is over the marbles.
    (Figure 21-2)

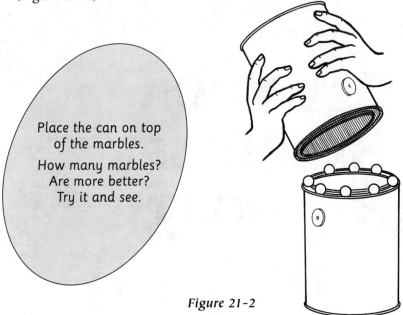

Place the can on top
of the marbles.

How many marbles?
Are more better?
Try it and see.

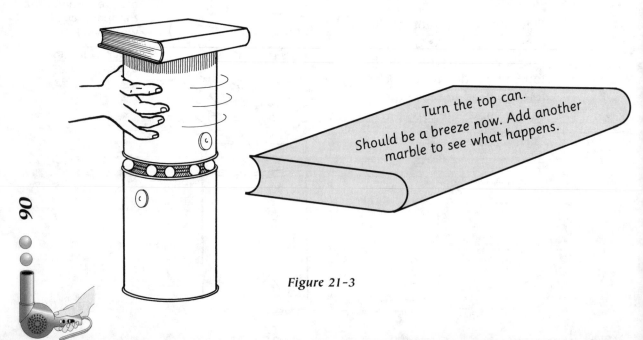

Figure 21-2

3   Place the books on top and notice how much effort is used to turn the
    top can. How do the two forces needed to turn the can compare?
    (Figure 21-3)

Turn the top can.
Should be a breeze now. Add another
marble to see what happens.

Figure 21-3

4 Rest a marble on a flat surface. Look at the point where the marble touches the surface. How does the size of this area affect friction? (Figure 21-4)

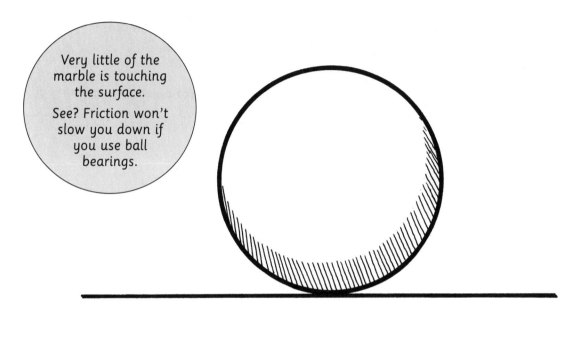

Very little of the marble is touching the surface.

See? Friction won't slow you down if you use ball bearings.

Figure 21-4

WHAT HAPPENED?

Ball bearings are hard and very little contact is made with the surface. Also, ball bearings are really rollers that are not limited in their direction of rotation because they are spheres. Rolling friction is much weaker than sliding friction.

Why do experimental solar-powered cars have narrow tires? Which do you think produces the least amount of friction, roller skates or inline skates?

91

GUESS WHAT?

★ Your bicycle wheels would not be able to turn without ball bearings. And you probably couldn't steer it, either.

★ Tiny ball bearings in watches are made from gems such as rubies.

Hang around for this project!

HOVERCRAFT

YOUR CHALLENGE

To observe how air can be used to reduce friction.

DO THIS

⚠ 1 Ask an adult to cut off the top part of the plastic bottle. Be sure to cut a smooth, even circle. Screw the cap in place. (Figure 22-1)

YOU NEED

Plastic soft drink bottle with cap

Knife

Hammer

Nail

Large balloon

Cut the top from the bottle. Finish your soda first!

2 Use the hammer and nail to punch a small hole in the center of the cap. (Figure 22-2)

Make a hole in the lid.

Turn the lid over to make it easier. Watch those fingers!

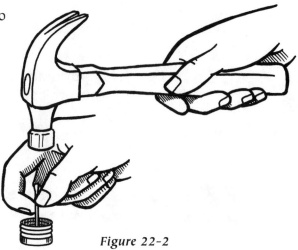

Figure 22-2

3 Blow up the balloon, twist the end to temporarily hold the air, and stretch the opening over the cap of the bottle. (Figure 22-3)

Attach the balloon to the lid.

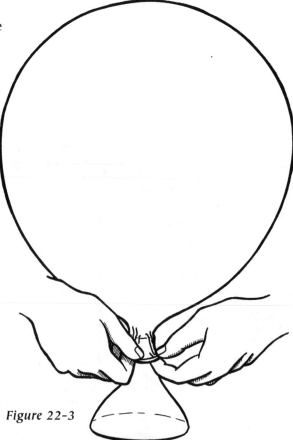

Figure 22-3

4 Release the balloon. What do you see? Why do you think this happens? (Figure 22-4)

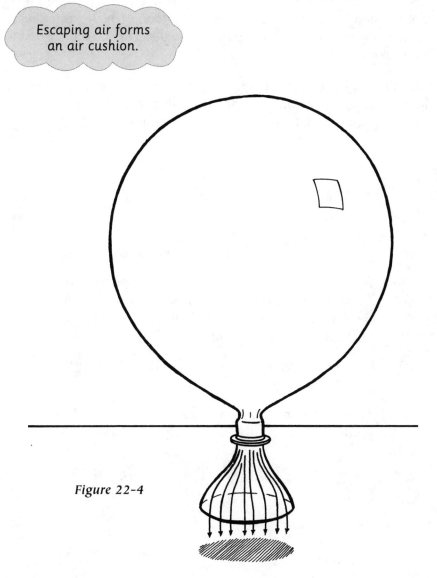

Escaping air forms an air cushion.

Figure 22-4

WHAT HAPPENED?

Air can be used as a lubricant to reduce friction. Hovercrafts are also called "air-cushion vehicles." They float on a cushion of air trapped inside a skirt. Air is made of tiny particles, and when these particles are trapped inside a container (like a balloon or hovercraft skirt), they fill the container and push against its walls. This pushing force is called *air pressure*.

The shape of a hovercraft is important. The bottom must have an area large enough for the air particles to push against and lift the craft. It is this air pressure that makes the hovercraft hover.

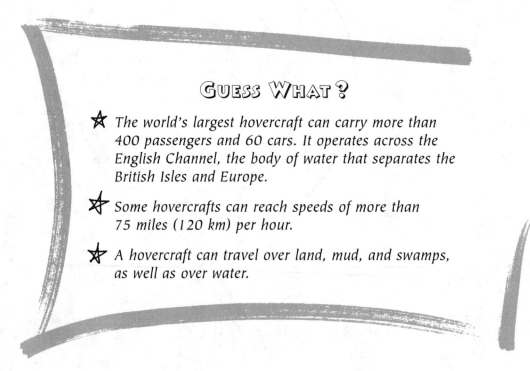

GUESS WHAT?

★ *The world's largest hovercraft can carry more than 400 passengers and 60 cars. It operates across the English Channel, the body of water that separates the British Isles and Europe.*

★ *Some hovercrafts can reach speeds of more than 75 miles (120 km) per hour.*

★ *A hovercraft can travel over land, mud, and swamps, as well as over water.*

You're getting
verrry sleepy...

TICK-TOCK

YOUR CHALLENGE

To investigate the mechanics of a simple pendulum.

DO THIS

1 Tie one end of the string to the weight and stretch the string out on a flat surface.

2 With a tape measure or ruler, measure the string to a point 39 inches (99 cm) from the weight and make a mark. (Figure 23-1)

Measure and mark the string.

Use white string so you can see the mark.

Figure 23-1

YOU NEED

Piece of string about 4 feet (120 cm) long

Tape measure

Weight (such as lead fishing sinker or nut from bolt)

Table or desk with the top at least 4 feet (120 cm) from floor

Felt-tip marker

Tape

Clock with second hand

97

3   Tape the free end of the string to the edge of the table so that the mark is at the bottom of the tape. This means that the distance from the weight to the bottom edge of the table is exactly 39 inches (99 cm) and the weight is free to swing in a wide arc. (Figure 23-2)

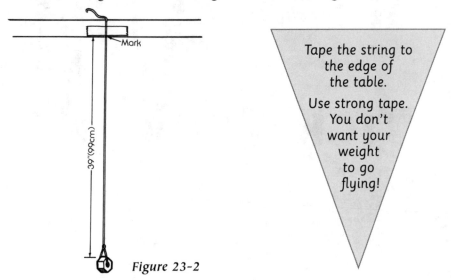

Mark

39"(99cm)

Tape the string to the edge of the table.

Use strong tape. You don't want your weight to go flying!

Figure 23-2

4   Place the clock behind the pendulum as it swings. Pull the weight to one side and release it as the second hand moves. Notice the motion of the weight. (Figure 23-3)

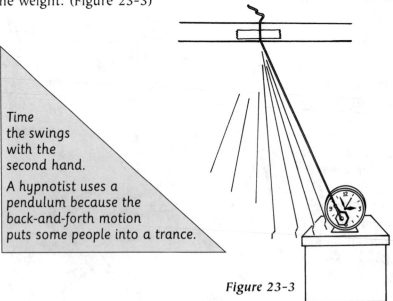

Time the swings with the second hand.

A hypnotist uses a pendulum because the back-and-forth motion puts some people into a trance.

Figure 23-3

What do you see? What makes the weight move? What would change if the weight were heavier or lighter? What happens if the string is shortened?

What Happened?

When you pulled the weight to one side, the energy you exerted to overcome gravity was stored in the weight as *potential energy*, or energy of position. When you released the weight, this energy became *kinetic energy*, or energy of motion. As the weight traveled to the end of its arc, it briefly paused. The kinetic energy changed back to potential energy for an instant, then back to kinetic energy as it moved. This change of energy is a continuous change as the pendulum falls from its maximum height to its minimum height (or its minimum speed to its maximum speed). This movement would continue forever if it were not for the friction and resistance of the air.

The time it takes for the weight to go out and back is called the pendulum's *period*. In this experiment the period should be 2 seconds. If the string was about 9.8 inches (25 cm) long, the period would be about 1 second.

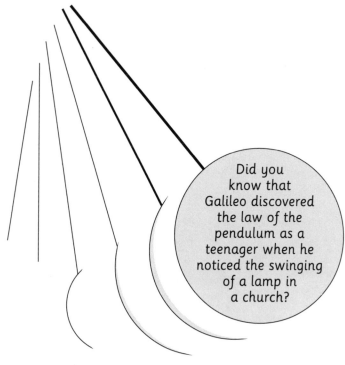

Did you know that Galileo discovered the law of the pendulum as a teenager when he noticed the swinging of a lamp in a church?

Guess What?

⭐ *A pendulum made with an iron ball suspended on a wire about 200 feet (61 meters) long was once used to prove that the earth rotated.*

⭐ *A special type of pendulum called a metronome is used to maintain a tempo for music. A metronome is actually an upside-down pendulum with a sliding weight to alter the period of the pendulum.*

You'll lose your marbles
with this project!

GIVE-AND-TAKE

YOUR CHALLENGE

To observe how the energy in a moving object is transferred
to stationary objects or objects at rest.

DO THIS

1 Lean the books against a wall with the bindings down,
   making a track on top for the marbles.

2 Prop one end of one of the books up about 1 inch to
   make a ramp. Keep the top level with the next book.
   (Figure 24-1)

YOU NEED

About eight marbles

**Two or three hardback
books about 1 inch
(2.5 cm) thick**

Make a grooved
ramp with books.

You'll need
hardback books
for this
experiment.

Figure 24-1

101

3 Place the marbles on the level part of the track. Have all of the marbles touching each other.

4 Now take one of the marbles up the ramp and release it. What do you observe? Can you explain what happens? (Fig. 24-2)

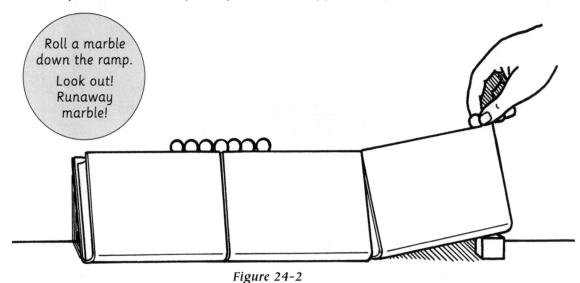

Roll a marble down the ramp.

Look out! Runaway marble!

Figure 24-2

5 Roll two or three marbles down the ramp at the same time. (Figure 24-3)

Roll two marbles together down the ramp.

Anyone who's ever played a game of marbles knows what happens next.

Figure 24-3

Where do the marbles get their energy? What happens to the energy of the marble rolling down the ramp? How does this compare with two marbles striking the row of stationary marbles?

WHAT HAPPENED?

When one marble strikes the row of marbles, one end marble will roll away. If two marbles are rolled down the ramp, two end marbles will roll away. If three marbles are rolled down the ramp, three of the end marbles will roll. This happens because the amount of energy of the marbles being rolled down the ramp is transferred equally to the marbles that are struck.

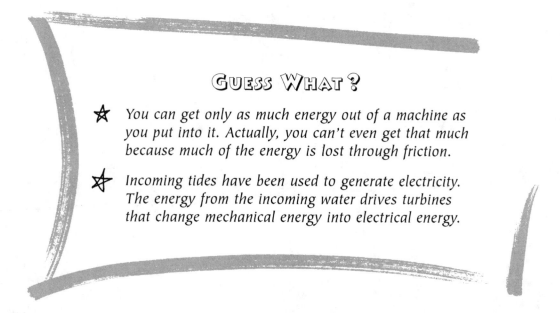

GUESS WHAT?

★ *You can get only as much energy out of a machine as you put into it. Actually, you can't even get that much because much of the energy is lost through friction.*

★ *Incoming tides have been used to generate electricity. The energy from the incoming water drives turbines that change mechanical energy into electrical energy.*

No one will force
you to do this!

FORCE OF NATURE

YOUR CHALLENGE

To investigate the forces necessary to move an object.

YOU NEED

Length of thread or light string

Brick

Short stick

DO THIS

1   Tie one end of the thread around the middle of the brick and tie the other end around the middle of the stick. (Figure 25-1)

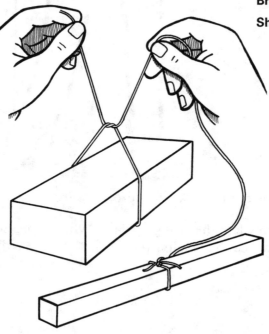

Tie the string around the brick.

Make sure to use very light string or thread.

2 Using the stick for a handle, slowly raise the brick a few inches from the ground. (Figure 25-2)

Figure 25-2

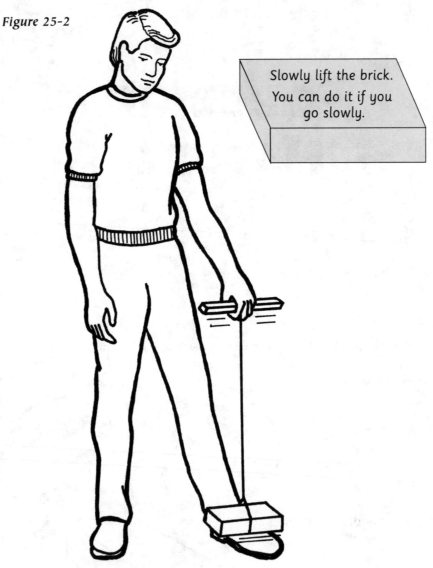

Slowly lift the brick.
You can do it if you
go slowly.

3 Lower the brick back to the ground.

4 Now jerk the string upward. What happens? What do you think caused the results? (Fig. 25-3)

Figure 25-3

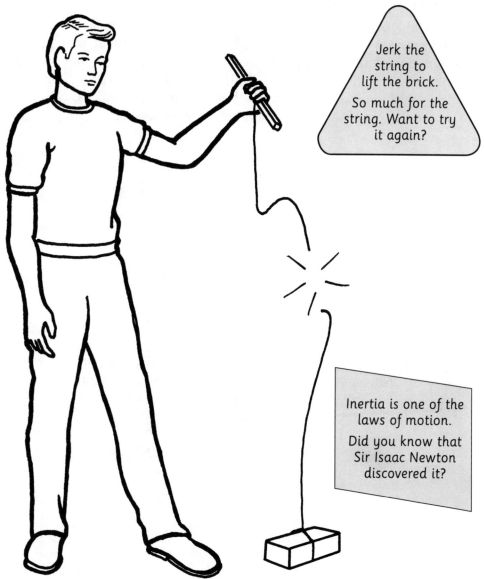

WHAT HAPPENED?

The *law of inertia* states that any body at rest tries to stay at rest, while a body in motion will move in a straight line until it is affected by some outside force. It takes a larger force to make a body at rest move abruptly than if it is done more gradually. It also takes a large force to stop a moving object quickly. The force with which an object moves, called *momentum*, is equal to its mass multiplied by its speed.

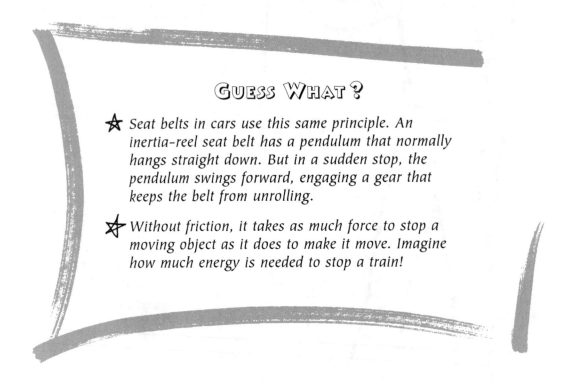

GUESS WHAT?

★ Seat belts in cars use this same principle. An inertia-reel seat belt has a pendulum that normally hangs straight down. But in a sudden stop, the pendulum swings forward, engaging a gear that keeps the belt from unrolling.

★ Without friction, it takes as much force to stop a moving object as it does to make it move. Imagine how much energy is needed to stop a train!

This one can turn around your
ideas about motion!

SPINNING CAN

YOUR CHALLENGE

To observe how one action produces an equal but opposite reaction.

DO THIS

 1 Using the hammer and nail, carefully punch about four holes of equal distance around the bottom of the can, or have an adult do it. As you remove the nail from each hole, push the nail to one side. Push the nail in the same direction each time, making all of the holes point in the same direction. (Figures 26-1 and 26-2)

YOU NEED

Hammer

Nail

Empty soda pop can

Length of string

Running water

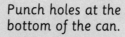
Punch holes at the bottom of the can.

To make this easier, put the can in a vise or have someone hold it. But be careful!

Figure 26-1

Figure 26-2

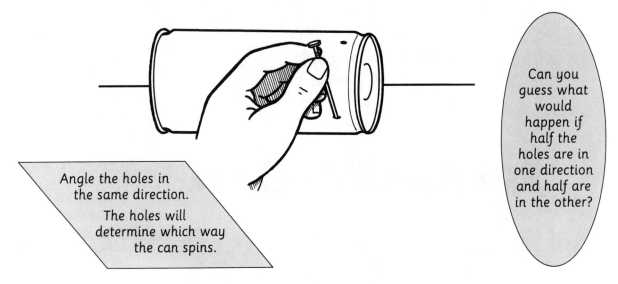

Angle the holes in the same direction.

The holes will determine which way the can spins.

Can you guess what would happen if half the holes are in one direction and half are in the other?

2 Bend the tab at the top of the can straight up and tie one end of the string in the tab opening. (Figure 26-3)

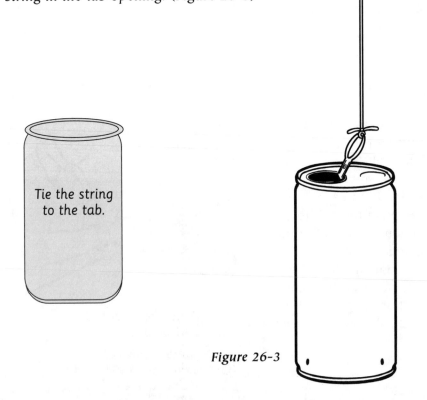

Tie the string to the tab.

Figure 26-3

3   With the other end of the string, lower the can under the kitchen faucet and fill it with water. Turn off the faucet.

4   Lift the can with the string. What do you observe? Why do you think this happens? (Figure 26-4)

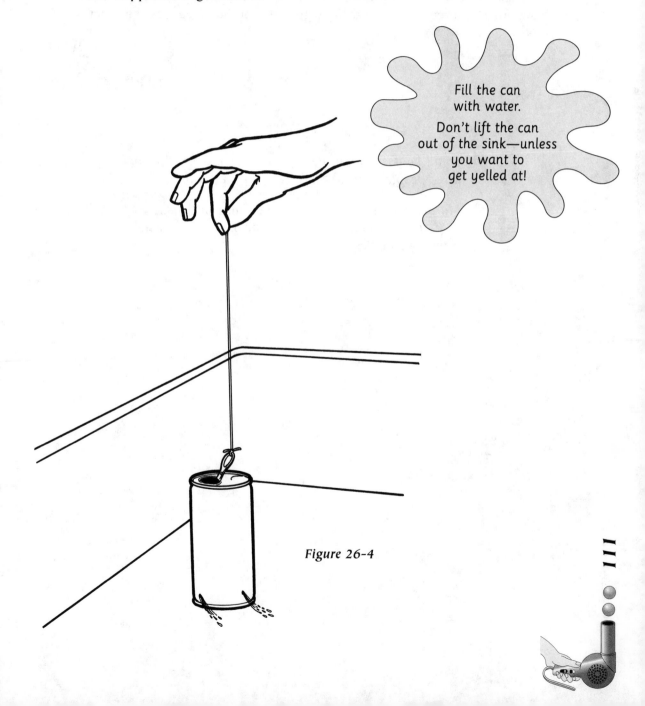

Fill the can with water.

Don't lift the can out of the sink—unless you want to get yelled at!

Figure 26-4

WHAT HAPPENED?

For every action, there is an equal but opposition reaction. In this experiment, the can is suspended by a string offering very little resistance. The water shoots out of the can at an angle and provides the action. The force of the water flowing from the holes causes the reaction.

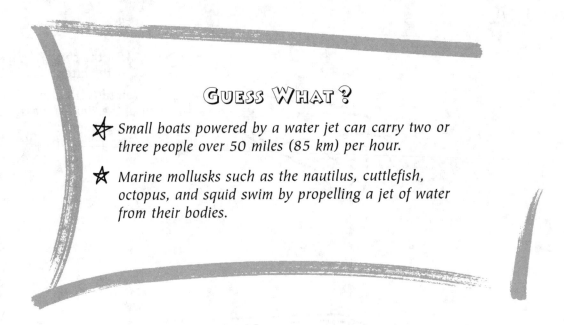

GUESS WHAT?

⭐ *Small boats powered by a water jet can carry two or three people over 50 miles (85 km) per hour.*

⭐ *Marine mollusks such as the nautilus, cuttlefish, octopus, and squid swim by propelling a jet of water from their bodies.*

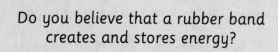

Do you believe that a rubber band creates and stores energy?

RUBBER BAND CAN

YOUR CHALLENGE

To investigate how energy can be stored and then reused.

DO THIS

⚠ 1 Using the hammer and nail, carefully make a hole about one-third of the way across the bottom of the can and another about two-thirds of the way. (Figure 27-1)

YOU NEED

Coffee can with lid

Hammer

Nail

Two rubber bands

Four toothpicks

Weight (fishing sinker or washer)

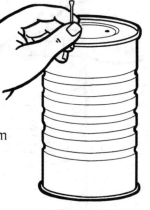

Figure 27-1

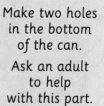

Make two holes in the bottom of the can.

Ask an adult to help with this part.

2 Now make similar holes in the lid.

3 Thread the end of one rubber band through one of the holes in the bottom of the can. Fasten this end with one of the toothpicks.

4 Thread one end of the other rubber band through the other hole in the bottom and fasten it with a toothpick. (Figure 27-2)

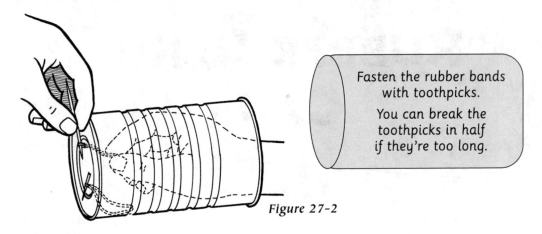

Fasten the rubber bands with toothpicks.

You can break the toothpicks in half if they're too long.

Figure 27-2

5 Thread the other ends of the two rubber bands through the holes in the lid and fasten them with the two remaining toothpicks. (Figure 27-3)

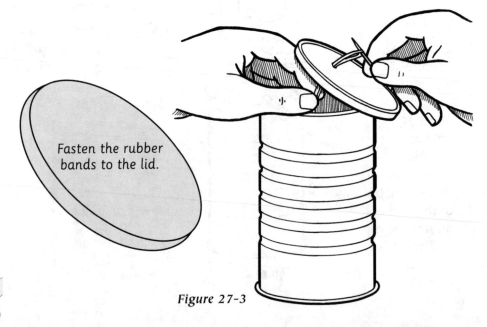

Fasten the rubber bands to the lid.

Figure 27-3

6 Tie the two rubber bands together halfway between the lid and the bottom of the can and attach the weight to this point. The rubber bands will form an X inside the can, with the weight tied to the center. Replace the lid. (Figure 27-4)

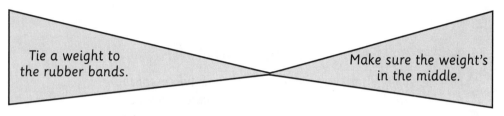

Tie a weight to the rubber bands.

Make sure the weight's in the middle.

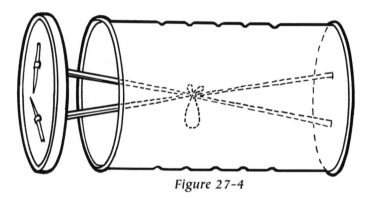

Figure 27-4

7 Roll the can on a smooth surface. What happens? Can you explain why? (You might want to try different sizes of rubber bands and weights to obtain better results.)

What Happened?

Rubber bands have *elasticity*. This means that a rubber band can stretch and twist under a force, and when the force is removed, it will almost return to its original shape. Inside the can, the weight and gravity create the force that twists the rubber band. The energy used to push the can will be stored in the twisted rubber band.

GUESS WHAT?

⭐ A flywheel *is a heavy wheel that, when spinning, can be used to store and deliver mechanical energy. A flywheel in a car is what makes the engine run smoothly.*

⭐ *Fossil fuels such as coal, oil, and natural gas contain energy from the sun stored from millions of years ago.*

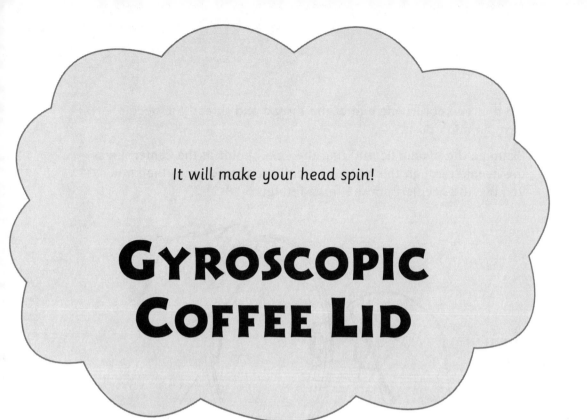

It will make your head spin!

GYROSCOPIC COFFEE LID

YOUR CHALLENGE

To observe the stabilizing effects of a gyroscope.

DO THIS

1   Place the pennies equal distances around the edge of the
    lid and fasten them in place with tape. They will add
    weight and balance to the lid. (Figure 28-1)

Fasten the
pennies with
tape.

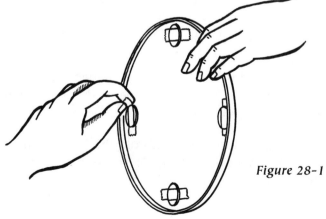

Figure 28-1

YOU NEED

Four pennies

**Plastic lid from
coffee can**

**Transparent or
masking tape**

**Length of thread
about 3 or 4 feet
(1.2 meters) long**

Sewing needle

2 Tie a large knot in one end of the thread and thread the needle, or have an adult do it.

3 Examine the plastic lid and find the small point at the center. Press the needle through this point and pull the thread up to the knot. Remove the needle from the thread. (Figure 28-2)

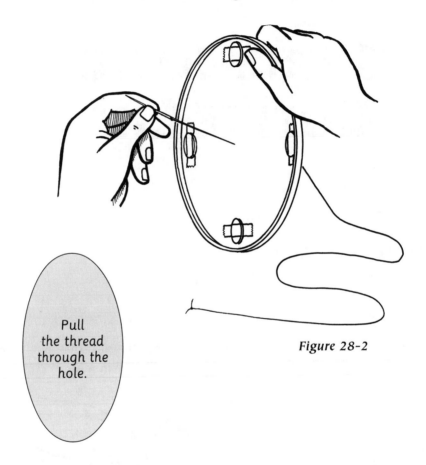

Pull the thread through the hole.

Figure 28-2

4 Hold the free end of the thread and suspend the lid a little above the floor. Swing the lid slowly from side to side like a pendulum. What do you observe? (Figure 28-3)

5 Now try it again, but this time, hold the lid level and give it a spin. What happens this time? Why do you think the lid behaves this way? (Figure 28-4)

Figure 28-3

Swing the lid back and forth.
Don't go into a trance!

Figure 28-4

Swing the spinning lid back and forth.

WHAT HAPPENED?

As long as the gyroscope is spinning, it will remain in the same plane, or *attitude*, it had when it began to spin. This is why a spinning top will stand on its point and why the rotating wheels of a bicycle keep it upright.

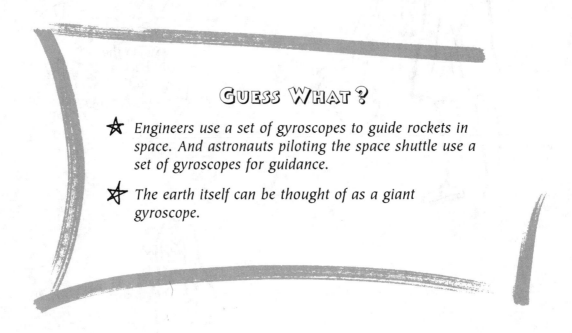

GUESS WHAT?

★ *Engineers use a set of gyroscopes to guide rockets in space. And astronauts piloting the space shuttle use a set of gyroscopes for guidance.*

★ *The earth itself can be thought of as a giant gyroscope.*

This project will have you going in circles!

May the Force Be with You

Your Challenge

To investigate the forces on an object in orbit.

Do This

1   Tie one end of the string securely around the tennis ball.
(Figure 29-1)

2   Find a clear place outdoors with no one standing nearby.

 3   Holding the other end of the string, start the ball swinging
in a circle around your head. Not too fast, just enough to
keep the ball and string level. (Figure 29-2)

As long as you hold the string and continue to swing, the ball
will travel in an orbit around your head. What forces do you
think are acting on the ball?

You Need

**4 feet (1.2 meters)
of strong string**

Tennis ball

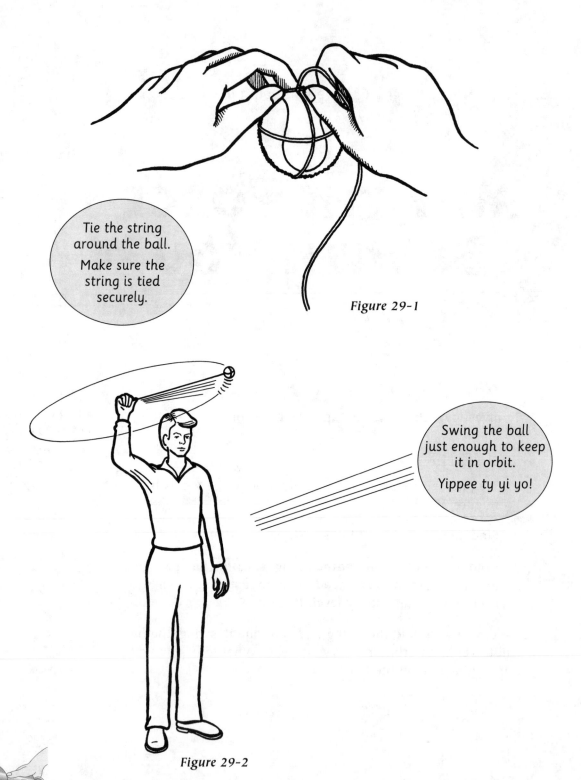

Figure 29-1

Figure 29-2

What Happened?

As the ball travels around, its momentum, or force of movement, pulls it *away* from the center of its orbit. This force is called *centrifugal force*. If the string breaks, the ball will fly away. This is caused by *inertia*, the principle that states that objects in motion tend to stay in motion. But the string is creating another force that tries to pull the ball in *toward* the center of the orbit. This is called *centripetal force*. It is balanced against the inertia and keeps the ball in orbit.

Guess What?

⭐ *Satellites in space are not floating. They are actually falling toward the earth. Because of their speed, around 18,000 miles (30,000 km) per hour, and their altitude, the earth's surface curves away at the same rate of fall, so the satellites never get any closer.*

⭐ *Television satellites orbit the earth above the equator at the same speed as the rotation of the earth. This means that they are always above the same point on the earth's surface. They are called* geostationary *satellites.*

⭐ *Other examples of centripetal force at work include playground merry-go-rounds and washing machines. After the wash cycle is completed, the clean, wet clothes are pressed up against the sides of the barrel because of centripetal force.*

You'll need to save your
energy for this project!

ENERGY
CONSERVATION

YOUR CHALLENGE

To observe how moving objects try to
maintain the same level of energy.

DO THIS

1   Thread one end of the string
    through the opening in the weight
    and tie securely with a knot.

2   Feed the other end
    of the string
    through the hole
    in the spool.
    (Figure 30-1)

 3   Find a clear place
    outdoors with no one
    standing nearby.

YOU NEED

**Strong string about
5 to 6 feet (1.5 to
1.8 meters) long**

**Weight (small fishing
sinker or 2 or
3 washers)**

Empty thread spool

Slide the string
through the spool.

4 Hold the free end of the string tightly in one hand, and hold the spool over your head with the other hand. Start the weight swinging in a large circle. (Figure 30-2)

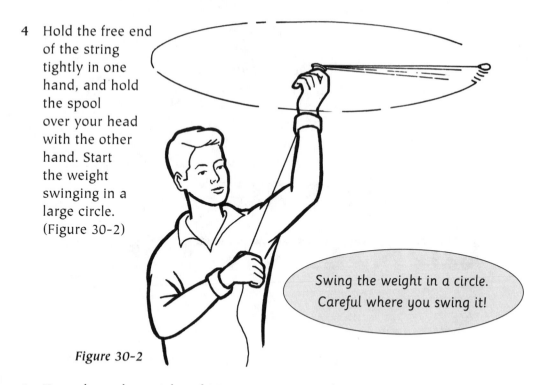

Swing the weight in a circle. Careful where you swing it!

Figure 30-2

5 Try to keep the weight orbiting at a nice steady rate. Notice the speed the weight is traveling. Hold the spool at the same height and pull down on the end of the string. What happens? (Figure 30-3)

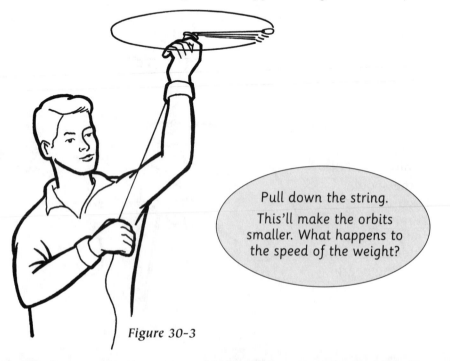

Pull down the string.
This'll make the orbits smaller. What happens to the speed of the weight?

Figure 30-3

WHAT HAPPENED?

When the weight was traveling in a larger orbit, it was moving at a steady speed with a certain amount of energy. Then the orbit became smaller. In order to keep the same energy level, it had to make more orbits, or revolutions, for the same amount of time. The weight had to increase its speed.

GUESS WHAT?

⭐ *Energy can never be created or destroyed. It can only be changed from one form to another.*

⭐ *Almost any time energy is converted from one form to another, heat is produced as wasted energy.*

Try this with different things around the house—just don't use your little brother!

FALLING BODIES

YOUR CHALLENGE

To investigate the laws of falling bodies.

DO THIS

1   Place the pillow on the floor.

2   Hold the book over the pillow in one hand and the sheet of paper in the other. Hold both level with the floor.

3   Drop both objects at the same time. Which one fell the fastest? (Figure 31-1)

4   Now place the sheet of paper flat on top of the book and repeat the test. Now what happens? Why do you think the paper does this? (Figure 31-2)

YOU NEED

Pillow

Book

Sheet of paper

129

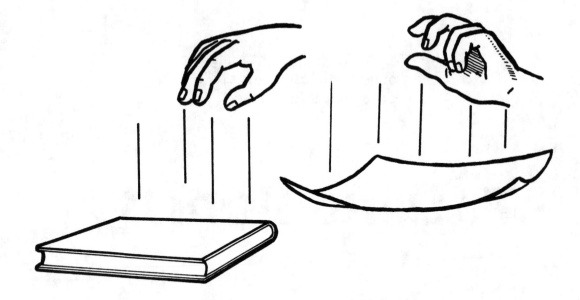

Drop the book and the paper at the same time.

Didn't Galileo do an experiment like this off the Leaning Tower of Pisa?

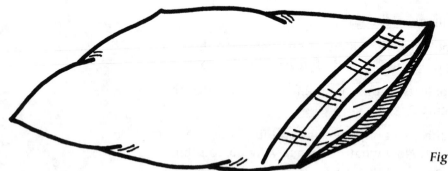

Figure 31-1

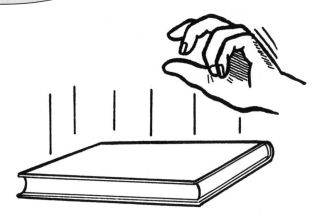

Drop the book with
the paper on top.

What happens to the rate of
the paper's fall?

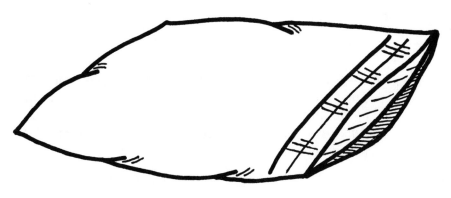

Figure 31-2

WHAT HAPPENED?

Without the resistance of air, gravity makes all falling objects drop at the same constant acceleration, regardless of how much they weigh. The acceleration of gravity is expressed in terms of the increase of velocity per second at a rate of about 32 feet (9.8 meters) per second, per second. This means that the velocity increases 32 feet (9.8 meters) per second for each second of fall.

For example, at the end of the *first* second, the book will fall at a velocity of 32 feet (9.8 meters) per second; at the end of the *second* second, it will fall at 64 feet (19.6 meters) per second; and at the end of the *third* second, it will fall at 64 (19.6 meters) plus 32 feet (9.8 meters) per second, or 96 feet (29.4 meters) per second. The book will fall faster and faster until the resistance of the air balances the pull of gravity. The book will then fall at a constant velocity, called the *terminal velocity*.

GUESS WHAT?

★ *A feather and a cannonball will fall at the same speed in a* vacuum, *or a space with no air.*

★ *Skydivers in a stable, horizontal position reach their terminal velocity of about 120 miles (190 km) per hour in about 8 seconds.*

★ *If a ball is thrown perfectly level with the ground at the same time another ball is dropped from the same height, the thrown ball will travel farther horizontally, but they both will hit the ground at the same time. The horizontal motion does not change the rate of fall of the thrown ball. It is traveling, but it is also falling. And it will fall at the same speed as the ball that was dropped.*

Time to change direction!

PUSH AND PULL

YOUR CHALLENGE

To observe how slightly changing the direction of a force affects a round object.

DO THIS

1 Unwind about 2 feet (60 cm) of thread and place the spool on its side on the table. Have the thread coming from the bottom of the spool.

2 Hold the end of the thread about 1 foot (30 cm) above the table and pull on the string. What happens? (Figure 32-1)

3 Now lower the end of the thread to about 1 inch (2.5 cm) above the table and try again. What happens this time? Why do you think this happens? (Figure 32-2)

133

Figure 32-1

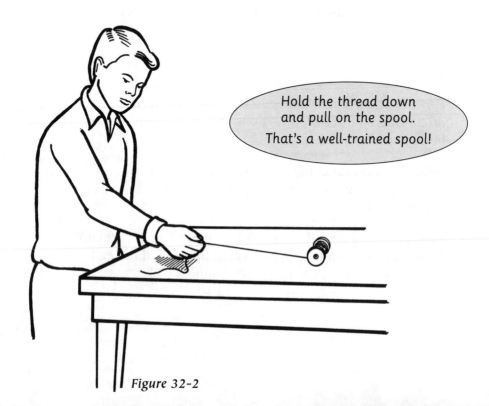

Figure 32-2

WHAT HAPPENED?

A force must be applied in the same direction you want an object to move.

GUESS WHAT?

⭐ *Nothing moves without some force. For example, astronauts in space cannot pour a drink into a glass because of the lack of gravity.*

⭐ *Because we are on the earth, we are traveling in an orbit around the sun at an average speed of about 66,000 miles (105,600 km) per hour.*

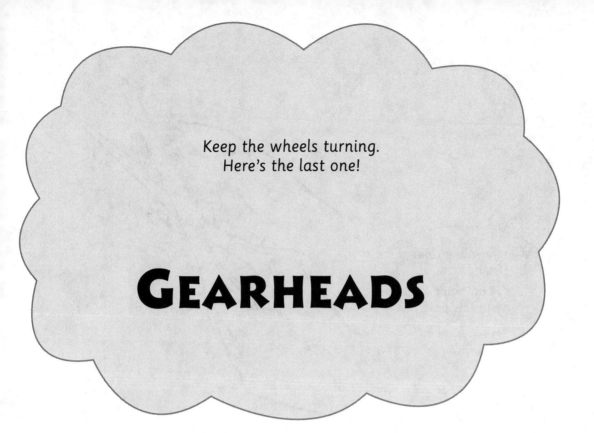

Keep the wheels turning.
Here's the last one!

GEARHEADS

YOUR CHALLENGE

To observe how gears are used to change the direction and
speed of a force.

DO THIS

1   Examine the plastic lids and notice the grooves cut in the
side. You will also see a small ridge on the bottom edge
of the cap.

 2   Using the scissors, trim off this ridge, or have an adult do
it, so that the sides of the lid are flat except for the
grooves. (Figure 33-1)

 3   Place one of the lids, top down, on the block of wood and
drive a nail through its center. Don't hammer the nail too
far, just enough to hold the lid in place.
(Figure 33-2)

Scissors

**Three plastic lids
from milk jugs**

Three small nails

**Block of wood
for base**

Hammer

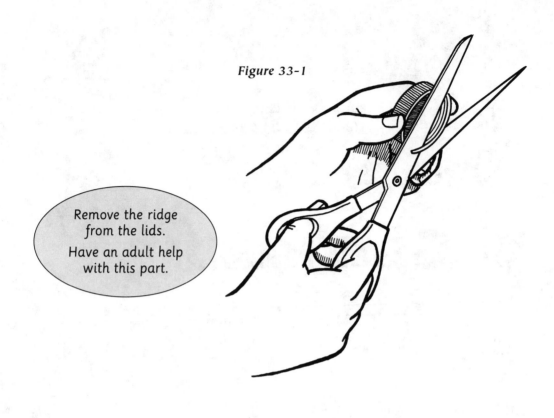

Figure 33-1

Remove the ridge from the lids.
Have an adult help with this part.

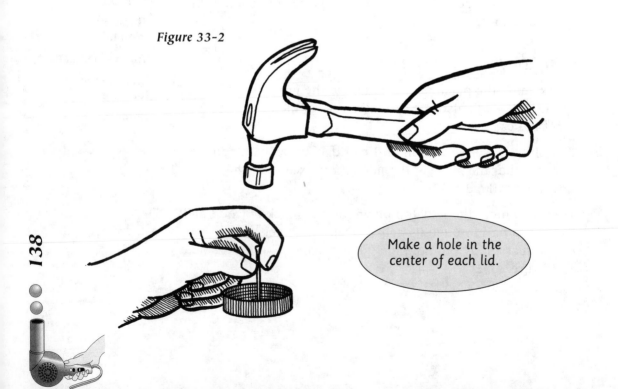

Figure 33-2

Make a hole in the center of each lid.

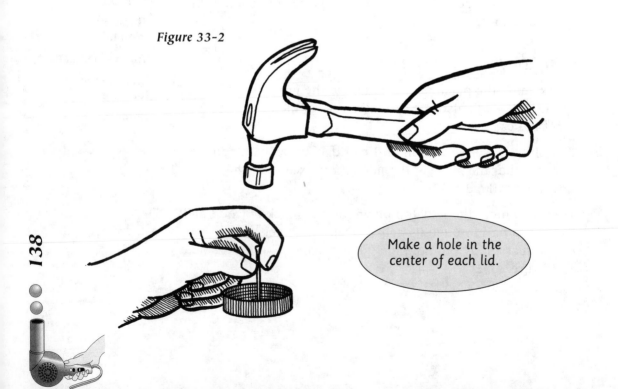

4   Now rotate the lid with your fingers until the lid turns freely. The nail becomes an axle and the lid a gear.

5   Place another lid against the first lid so that its grooves fit into the grooves of the first lid. Fasten it the same way.

6   Rotate the second lid until it also turns easily. What happens when you turn the second lid?

7   Repeat the steps with the third lid against the second. Turn one of the lids. What happens to the other lids? (Figure 33-3)

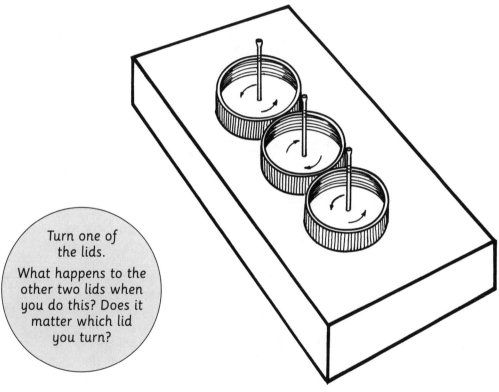

Turn one of the lids.

What happens to the other two lids when you do this? Does it matter which lid you turn?

Figure 33-3

8   Substitute smaller and larger plastic lids with the same-size grooves and try the experiment again. What happens when a small lid turns a larger one? How about when a large lid turns a small one?

WHAT HAPPENED?

A *gear* transfers rotating motion and power from one part of a machine to another. One of the most common gears is a metal wheel or disk with teeth cut into the edge. These teeth fit into the teeth of another gear so that when the first gear (called the *power gear*) turns, it causes the second gear to turn.

By using a small gear with a larger one, you can also change the force and speed of the turning gear. For example, if the power gear is 6 inches (15 cm) in circumference and the other gear is 12 inches (30 cm), the smaller gear will turn twice while the larger turns only once. The speed is cut in half, but the turning force, called the *torque* (pronounced "tork"), is doubled. If the power gear is the larger one, then the turning force is reduced, but the speed will be doubled.

How do gears change the power and speed of a bicycle?

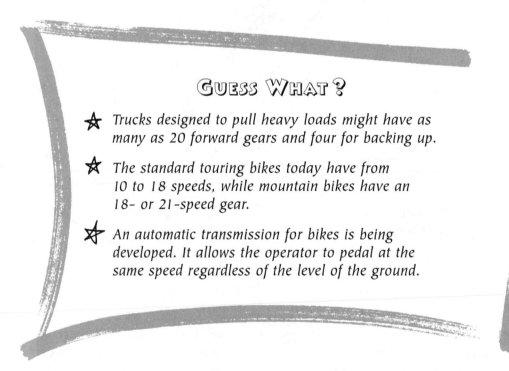

GUESS WHAT?

★ Trucks designed to pull heavy loads might have as many as 20 forward gears and four for backing up.

★ The standard touring bikes today have from 10 to 18 speeds, while mountain bikes have an 18- or 21-speed gear.

★ An automatic transmission for bikes is being developed. It allows the operator to pedal at the same speed regardless of the level of the ground.

GLOSSARY

acceleration The rate of change of velocity in a moving body.

aerodynamics The study of forces exerted by air or other gases in motion.

airfoil An object, such as an airplane wing, designed to keep an aircraft up or control its movements by reacting to the air around it a certain way. Also, a similar device on a car, such as a spoiler, or a boat, such as a sail.

centrifugal force A force tending to pull an object outward when it is rotating around a center.

centripetal force The force tending to pull an object toward the center of rotation when it is rotating around a center.

density The compactness of an object.

displace To move out of the proper position. For instance, water is *displaced* when a rock is dropped into it.

dynamics The branch of mechanics dealing with the motion of objects under the action of given forces; kinetics.

elasticity The condition of being able to spring back to its original size, shape, or position after being stretched or flexed.

electromagnetism A force that travels in waves through space that include radio, visible light, and x-ray waves. The branch of physics that studies electricity and magnetism.

friction The resistance to motion of two moving objects or surfaces that touch.

fulcrum The support or point of support on which a lever turns in raising or moving an object.

hydraulics The branch of physics that deals with liquids in motion.

hydrostatics The branch of physics having to do with the pressure and stability of water and other liquids; statics of liquids.

inertia The tendency of an object to remain at rest if at rest, or, if moving, to keep moving in the same direction, unless affected by some outside force.

kinetic energy The energy of an object that is created from its motion.

mass The amount of matter in a body as measured by its inertia. *Weight*, on the other hand, is a measure of the gravitational force on the body.

momentum The mass of a moving body multiplied by its velocity; the quantity of the motion of a moving body.

potential energy Energy that is the result of relative position instead of motion, as in a compressed spring.

specific gravity The ratio of the density of an object to the density of water.

static energy Energy not moving or at rest.

statics The branch of mechanics dealing with bodies, masses, or forces at rest or in balance.

surface tension The pull of any liquid on its open surface so that the surface is as small as possible. The resulting concentration of molecules form a thin skin.

velocity The rate of motion of an object in a particular direction.

INDEX

145

ABOUT THE GUY WHO WROTE THIS BOOK

A keen observer of nature and an avid follower of scientific advances, author Robert W. Wood injects his own special brand of fun into children's physics. His *Physics for Kids* series has been through 13 printings, and he has written more than a dozen other science books. His innovative work has been featured in major newspapers and magazines.